CULTURE, URBANISM

T0227915

Heritage, Culture and Identity

Series Editor: Brian Graham,
School of Environmental Sciences, University of Ulster, UK

Other titles in this series

Ireland's Heritages
Critical Perspectives on Memory and Identity
Edited by Mark McCarthy
ISBN 0 7546 4012 4

Senses of Place: Senses of Time
Edited by G. J. Ashworth and Brian Graham
ISBN 0 7546 4189 9

(Dis)Placing Empire
Renegotiating British Colonial Geographies
Edited by Lindsay J. Proudfoot and Michael M. Roche
ISBN 0 7546 4213 5

Preservation, Tourism and Nationalism
The Jewel of the German Past
Joshua Hagen
ISBN 0 7546 4324 7

Tradition, Culture and Development in Africa
Historical Lessons for Modern Development Planning
Ambe J. Njoh
ISBN 0 7546 4884 2

Culture, Urbanism and Planning

Edited by

JAVIER MONCLÚS and MANUEL GUÀRDIA
Polytechnic University of Catalonia, Spain

LONDON AND NEW YORK

First published 2006 by Ashgate Publishing

Published 2016 by Routledge
2 Park Square, Milton Park, Abingdon, Oxfordshire OX14 4RN
711 Third Avenue, New York, NY 10017, USA

First issued in paperback 2016

Routledge is an imprint of the Taylor & Francis Group, an informa business

Copyright © Javier Monclús and Manuel Guàrdia 2006

Javier Monclús and Manuel Guàrdia have asserted their moral right under the Copyright, Designs and Patents Act, 1988, to be identified as the editors of this work.

All rights reserved. No part of this book may be reprinted or reproduced or utilised in any form or by any electronic, mechanical, or other means, now known or hereafter invented, including photocopying and recording, or in any information storage or retrieval system, without permission in writing from the publishers.

Notice:
Product or corporate names may be trademarks or registered trademarks, and are used only for identification and explanation without intent to infringe.

British Library Cataloguing in Publication Data
Culture, urbanism and planning. – (Heritage, culture and
    identity)
    1. City Planning 2. Cities and towns – Historiography
    I. Monclus, F. J. (Francisco Javier) II. Guardia i Bassols,
    Manuel, 1949–
    307.1'216

Library of Congress Cataloging-in-Publication Data
Culture, urbanism and planning / edited by Javier Monclus and Manuel Guardia.
        p. cm. -- (Heritage, culture and identity)
    Includes index.
    ISBN: 0-7546-4623-8
    1. City planning--History. 2. City planning--Social aspects. 3. Cities and towns.
    4. Culture. I. Monclus, F. J. (Francisco Javier) II. Guardia i Bassols, Manuel, 1949– III.
Series.

    HT166.C835 2006
    307.1'21609--dc22

                                                                                            2006007774
ISBN 13: 978-1-138-25357-5 (pbk)
ISBN 13: 978-0-7546-4623-5 (hbk)

Contents

List of Figures

List of Tables

Foreword

Roque Gistau Gistau
President of Expo Zaragoza 2008

Javier Monclús and Manuel Guàrdia have edited an extensive book entitled *Culture, Urbanism and Planning*, analysing the interactions between these three concepts, which are responsible for the character and personality of a city, and sometimes even its soul and charm.The chapters all expand upon contributions made by authors in the context of the International Planning History Society Conference, which took place in Barcelona in 2004, under the theme of *Planning Models and the Culture of Cities*.

Beyond the narcissistic logic of consumption or the urban spectacularisation associated with numerous urban interventions over the last few years, a new form of urbanism has become possible and necessary, focusing on the cultural dimension. Strategic plans and projects adapted to particular, changing and uncertain contexts are becoming increasingly important. This 'new urbanism' is based on a greater degree of sensitivity to the environment and attention to urban landscapes that are undergoing profound changes.

Just as other cities that have entered a new phase of modernity, Zaragoza has been undergoing a substantial transformation over the last twenty years, both in terms of its economic and demographic base, and its social relations and cultural guidelines. Since General Motors set up just 25 km from the city centre in the early 1980s, the urban economy has been experiencing an intense process of diversification. Nowadays, Zaragoza no longer depends exclusively on the automobile and related industries and the city is committed to logistics and other emerging industries. This reorientation of activities, on which recent urban growth is based, is revealed by the arrival of the high speed train service (AVE), the construction of the Pla-Za logistics platform, the recycling industrial group and many other projects.

There has also been a change in the scale of new urban infrastructure and amenities. This new scale has led to a significant change in the role of public spaces, green corridors and in particular the watercourses that should form the backbone of this city: the rivers *Ebro, Huerva* and *Gállego* and the *Canal Imperial de Aragón*. The city has doubled its population over the last 50 years and is currently experiencing a different type of growth; around 10% of its 660,000 inhabitants are immigrants.

The International Exhibition dedicated to 'Water and Sustainable Development' will be held in Zaragoza in 2008, promoted by the society of which I am Chairman. Hosting Expo 2008 is an opportunity to boost growth even further, by working to achieve energy efficiency and the responsible use of water resources, thus making growth compatible with environmental sustainability. Furthermore, Expo will entail the construction of an important cultural, scientific and recreational park, conceived as a new focus of urban centrality capable of promoting the integration of the river Ebro in the city. Post Expo exploitation is therefore a strategic approach that must act as a catalyst for this metropolitan park, providing other plans and projects that are essential for the city.

The actions linked to Expo can be inscribed within this new form of urbanism that is more interested in culture and strategic projects than in traditional zoning. It is a set of exemplary actions to create new significant places, which aim to improve the quality of public spaces. The system of green corridors and the river park itself, in constant interaction with the river and river life, will be a permanent reminder of the importance of river ecosystems for environmental sustainability.

But beyond these actions, Expo 2008 should also be the seed of a new urban and environmental culture, incorporating criteria of quality, efficiency and sustainability. The challenge is to integrate strategic, urban and cultural visions. In other words, in the case of Zaragoza, such actions would not only seek to enhance the city's international presence and improve its citizens' quality of life, but also aim to make the city an international point of reference, creating a Zaragoza brand based on the responsible use of a unique resource such as water at the start of the twenty-first century.

May 2006

Introduction

F. Javier Monclús and Manuel Guàrdia

The loss of the habitual productive functions, that had supported the contemporary city up until the 1970s, has come to give 'culture' a leading role that in fact it had never completely lost. The material production was a secondary aspect during millenniums of existence of the cities, and the current situation could be seen as a simple return to normality if the word culture were to remit to a clear universe without ambiguities. The word, although irreplaceable, turns out to be uncomfortable for its polysemical and ambiguous characteristics. Its meaning is, on the one hand, impregnated historically by the autonomous, high and ethnocentric *enlightened* notion of culture. However at the same time it incorporates, from romanticism, a more comprehensive vision, that has culminated finally in anthropological approaches. The very economic thought adopted it, early on, against the abstract individualism of liberal economic theory.[1] In this line, during the nineteenth century, teaching, museums, forms of production and consumption, the celebration of large events ..., were conceived as strategic investments for encouraging the domestic economy. The large International Expositions or the Arts and Crafts movement, and their efforts to improve public taste, are in this sense exemplary episodes. Nevertheless, culture has enjoyed an ambiguous condition. It is considered that an advanced economy could not leave aside, but its values have remained outside the economic calculation and, in general, they have been situated in a superior sphere to the values of the ones occupied by the economy. Up until the 1960s, the common use of the word continued to be marked by the *enlightened* notion of high culture and was associated fundamentally to minority manifestations of an elitist character.[2] Over more recent few decades, the influence of anthropological thought, the decolonising process, the progress of mass culture, the contributions from critical theory and the increasing cultural diversity in

---

1 Friedrich List (1789–1846), precursor of the historicist school, highlighted the fact that economic practice revealed how the productive forces of cities were conditioned by laboriousness, the urge to save, the morality and intelligence of individuals, as well as the institutions, social, political and civil laws; at the time that the manufacturing energy encouraged science, art and political perfection.

2 In the sense resumed by Matthew Arnold's (1822–1888) phrase: 'the best that has been thought and said in the world'.

the cities themselves, have brought about change and a notion closer to that utilised by anthropology has become prevalent widening its meaning in a decided form.[3]

This less ethnocentric and more democratic vision of the term was associated, in the 1970s and at the start of the 1980s, to cultural politics implicated in social development. Since the 1980s there has been a double inflection. On the one hand, with the progressive loss of weight of physical production in cities, cultural politics have been converted into tools of economic development. Culture has become converted into premium material of the urban economy and, although it has persistently demanded an exceptional statute, it has been seen to be invaded by economic logic and the practices of *marketing*. In parallel, the vision over its social dimension has changed as a result of a growing awareness with respect to cultural diversity and place identity. At least, this is what is demanded from the more socially compromised approaches proposed by the renewal of urbanism.[4] Therefore it is possible to speak of a new prominence and of decisive inflection, but it should not be forgotten that in economic and urban development related thought, born with the tensions of the city in the industrial age, the cultural dimension was present right from the very beginning.

The presence of the terms urbanism and planning in the title, although redundant, tries to delimit a more comprehensive and fluid area of problems, than that which each term evokes on its own. It suggests the need to integrate visions that have developed independently from one another, often, as conflicting traditions. Michael Hebbert characterises them in a synthetic and controversial form:

> Planning is rooted in social reformism, is mainly Anglo Saxon and sees itself as a professional activity distinct from architecture and engineering. Urbanism is a shared culture or common ground between these professions, is Latin and owes more to the pluralism of the real urban politics.[5]

Nevertheless, one should not forget that the term planning and its different adjectives (town planning, urban planning and city planning), in the Anglo-Saxon world, as much as their Latin equivalents (urbanismo, urbanisme, urbanística) arose around the first decade of the twentieth century, to designate the activity controlling urban development, and in fact have a lot in common in their genealogy.

The term *urbanism* is often utilised in a broad sense as the 'study of cities – their economics, politics, social and cultural dimensions', and its early use in the United

---

3   The recent definition proportioned by UNESCO serves as an example: the 'set of distinctive spiritual, material, intellectual and emotional features of society or a social group and that it encompasses, in addition to art and literature, lifestyles ways of living together, value systems, traditions and beliefs' (UNESCO 2002).

4   D. Hayden, *The Power of the Place Urban Landscapes as Public History* (Cambridge, Mass.: MIT Press, 1995); S. Zutkin, *The Cultures of Cities* (Cambridge, Mass.: Blackwell, 1995).

5   M. Hebbert, 'Town Planning versus Urbanismo', in *11th Conference of the International Planning History Society (IPHS), Planning Models and the Culture of Cities* (Barcelona: UPC-ETSAV, 2005), p. 89 (revised version in Planning Perspectives). His approach is based on a quotation of Giorgio Piccinato.

States, is especially significant, in the sense of urban culture: the 'urbanism as a way of life' (Wirth, 1938). A meaning that would acquire a renewed importance and which was situated, in fact, in the origins of urban anthropology. Lewis Mumford's analyses, between the 1930s and 1960s, adopted this perspective contrasting it to the functionalist vision ever increasingly more dominant of the CIAM.[6] Neither is it strange that the critical visions from Jane Jacobs to Henri Lefebvre were aimed at this more social and cultural dimension of urbanism, and that a certain confluence would be produced between these visions and the more architectural ones, that in turn would result in the break with the planning tradition and the claim to the 'culturalist' urban project.[7]

At present, strategic planning and the large urban projects, motors of re-launching and renewal of the post-Fordist cities, have led to a reconsideration of the economic, social and cultural dimensions of urbanism. It is ever more difficult to dissociate these factors in urban development activity, which seems to demand renewed attention being paid to the modalities of planning, understood as a system of practices and as a process, more than as an activity centred on traditional *zoning* or in *urban* architecture.[8] The diversity of traditions and uses of both words thus depicts a widened field that is important to reconsider today.

The first three chapters offer a good frame of reference for setting out the relations between planning, urbanism, history and culture. Those relations were at the centre of Françoise Choay's influential analysis published in 1965, which marked the ascent of a new culturalist vision that has not ceased from developing and from acquiring prominence.[9] In that essay, Choay contrasted the progressive vision with the culturalist vision in the formative phase of urban development thought – the two fundamental positions according to the two perspectives (past and future) at that time. On the one hand the progressive vision looked to the future, exercising as an agent of modernisation. His critical base was the human individual and the universal reason, capable of resolving the problematic relation of man with the world through science and technology. On the other hand, culturalist planning looked to the past. It assumed the forms of nostalgia, and its critical base was the urban community and the assertion of the values of the historic city over modern banality. In this reading, the two models were presented as recurrent, in the different phases of the development

---

6   L. Mumford, *The Culture of Cities* (1938), (New York: Harvest, 1970); *The City in History* (New York: Harcourt, Brace and World, 1961).

7   M. Hebbert, 91: 'from 1970 onwards urbanism meant a return to the urban spaces of street and square'.

8   As Rem Koolhaas indicated 'If there is to be a 'new urbanism' it will not be based on the twin fantasies of order and omnipotence; it will be the staging of uncertainty; it will not be concerned with the arrangement of more or less permanent objects but with the irrigation of territories with potential; it will no longer aim for stable configurations but for the creation of enabling fields that accommodate processes that refuses to be crystallised into definitive form'. R. Koolhaas, 'What Ever Happened to Urbanism?', in *S, M, L, XL* (New York: The Monacelli Press, 1995), p. 969.

9   F. Choay, *Urbanismo: Utopías y realidades* (Barcelona: Lumen, 1970, 1965 1st).

of town planning. Nevertheless, considering the present 'cultural turn' as a simple reissue of this historic recurrence would be a clearly reductive interpretation.[10]

Michael Hebbert and Wolfgang Sonne adopt a more pragmatic than nostalgic attitude in their chapter, to analyse the uses of history in twentieth century planning. Of the three basic *patterns of reading the past* that they recognise, they emphasise the study of the *Lasting Types and Forms* and the use of *History as Local Memory* for being the most operative. They distrust the great interpretations, *the history like a Big Canvas*, and defend history for its usefulness in interpreting urban processes and helping new planning proposals. When the relations between planning and culture are examined, as occurs in the two subsequent chapters, it can be seen how today the use of the very word culture embraces a much wider field than the notion of culturalist urbanism, understood as the discipline looking towards the past. Currently the cultural economy integrates all kinds of artistic forms and institutions, including not only the media, but also publicity and marketing. This is the present field of reference of the new cultural urbanism that, at the same time, must face the challenges and conflicts rising out of the new multiculturalism.

The changing relations between culture and urbanism are analysed in the chapter by Robert Freestone and Chris Gibson and synthesised in an historic succession of paradigms in order to recognise the new aspects brought to each phase. They underline the growing prominence of culture, up to the complete fusion in 'the creative city'. Greg Young's article, although recognising the growing inclusion of the dimensions of culture 'across the entire spectrum of planning experience', advocates a reflection that aims towards a more effective 'integration of culture into planning'. In spite of the fact that it takes the ethical component of culture into consideration, it rejects the return to universal values of the Enlightenment and settles for 'openness and diversity', and the 'cross-fertilisation of collaborative post-modern approaches and insights'. In a purposeful and instrumental tone, the author presents the experiences of 'cultural mapping as structures, techniques positioned to capitalise on the whole culture of place, tangible and intangible elements, and to articulate cultural diversity and values'. Both chapters demonstrate the present notion of culture: multiple, dynamic, fluid and with changing hierarchies. They are a good exponent of how the looks towards very diverse, present, hidden or emerging cultural issues have multiplied.

Initially the conservation of monuments, as well as the formation of museums, was inscribed within a policy of 'construction of the national memory', a heraldic and emphasised history. Urbanism and the cultural facilities of capital cities were

---

10 F. Choay herself indicated the limitations of both positions: the ingenuous and persistent illusion of a scientific planning in the progressive approaches, as much as the blockade of the culturalist urbanism within an aesthetic outlook. Both models *a priori*, objectified and treated as things, ignored the fact that the built environment had the specific quality to be meaningful, and had exercised a corrosive influence over the imaginary. Choay advocated proposals *a posteriori* born out of the point of view of the inhabitants, from their existential, practical and affective links.

the modern shop window of a reconstructed identity, and the materialisation of the national spirit. The careful restoration of some monumental architecture in the renovated Athens, or in Haussmann and Viollet-le-Duc's Paris, was accompanied by the demolition of old urban fabrics. David Gordon's chapter shows how the weight of political matters and capital status continued exercising a decisive consideration throughout the twentieth century, and does not seem to be affected by the weakening of the national narrative and of the state frameworks. Nevertheless the social imaginary of the twentieth century was not formed in a top-down manner alone, at the dictates of political power and technical elites. Margarita Gutman shows the influence of images diffused through the most popular mass media, in Buenos Aires in the 1920s, reflecting the bottom-up approach.

Similarly the evolution of the conservationist sensibility ended up overflowing time and again the technical and institutional approaches. The Italian experience, expressed through Giorgio Piccinato's useful revision, offers the key principles in order to understand this process and its influence in our critical approaches.[11] The chapters by Laura Kolbe and John Pendlebury allow for the perception, beyond the significant differences between urban, cultural and institutional realities, of the chronological correspondence and the basic tuning of the different experiences. In an extremely similar process to that which occurred towards the end of the nineteenth century, following the grand urban reforms of the European capitals, the 'conservation movement' arose like a reaction to the demolitions deriving from the politics of modernisation of the post-Second World War period. These often confronted the technocratic criteria of politicians and technicians, with the formation of artistic and intellectual minorities, partisan to conservation, in tune with the new movements which began growing from 1968 onwards. The process of integrating a wide notion of culture in planning is not in the slightest way exempt from conflicts. Over the last decade, planning has been increasingly invaded by economic logic and affected by the powerful social changes deriving from globalisation. The growth of tourism and the multicultural character of historic centres configure a new and changing reality that requires new answers, as indicated by Alessandro Scarnato, in examining the cases of Florence and Barcelona. In short, everything leads one towards a growing conscience of the necessary renewal of one's key interpretation. This is even more important when the success of the post-modern cities is increasingly associated to the ability of their urban centres to propose themselves as 'the synthesis of the positive aspects of the whole city'.[12]

Insofar as the cultural and symbolic economy came to prevail in the 1970s, firstly in the North American cities, and then in other western metropolises, conventional

---

11   It is sufficient to recall names such as Camillo Boito, Luca Beltrami, Gustavo Giovannoni, Roberto Pane, Cesare Brandi, Saverio Muratore, Guido Canella, and episodes of great influence such as Padova's morpho-typological analysis or Bologna's plan of the 1960s and 1970s.

12   G. Amendola, *La ciudad postmoderna. Magia y miedo en la metrópolis contemporánea* (Madrid: Celeste Ed., 2000), p. 33.

planning gave way to a cultural and strategic urbanism, that saw a resource and a motor for urban changes within culture. It was not a question of the revaluation of the city's history and its architectural heritage, but rather of the 'selling' of the whole city and its different urban fragments. In a context reflecting the growing intensity of urban competition in the global market, cities committed themselves to the renewal of their images as a decisive factor according to the strategic objectives set out. The integration of the economic and cultural dimensions – shown in the prominence of the new notions of cultural urbanism and *strategic urban planning* – implies that the clear dichotomies established by Choay in the 1960s have been overcome. The convergence between culturalist urbanism and that more attentive to economic concerns and the driving of the urban processes was able to be observed. In the same way, the renewed role of image in recent urban development strategies cannot be understood simply as 'formalism' or 'historicism', but rather as a manifestation of the pragmatism that dominates recent politics. The reconfiguration process of the city as a system of scenographies has expressions, so characteristic of the post-modern cities, such as the thematised historic areas, the new *shopping malls*, or certain residential environments of 'neo-traditional' architecture, such as the 'successful' *New Urbanism*: the current expression closest to culturalist urbanism, understood as a nostalgic look to the past and critical of the present. Christopher Silver reminds us that, despite its integrative approaches and its use of history as a critical tool, New Urbanism has not taken on the challenge of resolving the large problems of cities and, in fact, has contributed to *sprawl* and segregation.

Without any doubt, the *par excellence* expression of the current pragmatism is in the proliferation of 'emblematic' projects, with spectacular architecture, and the thematisation of the city, in strategies directed towards promoting the image of the cities' labels (*City Branding*), with the consequent risk of banalisation of urban spaces. As indicated by Graeme Evans ('Branding the City of Culture: the Death of City Planning?'), it is a matter of the 'explicit branding of whole cities', in the way of the large corporations, that has its logic and its costs, and can produce 'uneven, unpredictable results'. This same context drives the strategic utilisation of international expositions, Olympic Games and other events that, despite a long tradition, has acquired a renewed interest with the objective of renewing and re-launching the image of the city, to mobilise the diverse agents and resources, and to catalyse new urban projects. As F. Javier Monclús emphasises, these strategic planning projects have no reason to be limited to prestige or 'emblematic' projects alone, as they allow for the encouragement of other urban regeneration-related developments and, in fact, 'are just as susceptible to be converted into mere policies of image, as in others generating considerable economic and social benefits'. With all of these, the examination carried out by Lilian Vaz and Paola Berenstein of a number of Latin American experiences, where the excesses of 'heritagisation, musealisation and disneyfication' are associated to the progressive gentrification of the historic centres, warns of the risk over the imposition of spectacularisation urbanism as the 'single model'. The review carried out by Roberto Segre of recent urban development and architectural operations in a range of Latin American cities has the intention of valuing them according to the diversity of the situations and of the results. It therefore offers an uneven balance, full of contrasts, but also of opportunities.

In this series of approaches, the analysis offered by Stephen Ward relating to the case of Baltimore offers a good reason for a final reflection. During the last few decades, Baltimore has been converted into a model, a type of 'touristic and cultural form of urbanism' that has inspired and continues inspiring numerous regeneration operations of *waterfronts* and urban areas. Nevertheless, the distance between that limited by its effects in the urban economy, and in the well-being of the whole population – and not only for a sector of users and visitors of the cities – and the overwhelming international success of the model turns out to be somewhat disturbing. Although it does not signify the discrediting of other initiatives that seek to provide an answer to the intense processes of urban reconversion associated, one way or another, to the logics of globalisation.

*   *   *

The idea for this collection of essays arose out of the 11th Conference of the International Planning History Society (IPHS), held in Barcelona in July 2004.[13] The interest expressed in the connection between *culture* and *planning*, taking into consideration the different meanings of this latter term, was present in a number of the debates and discussions which took place during that Conference. For this reason, it seemed interesting to prolong this discussion in the form of this collection of essays, to enable their diffusion for a much wider audience. These essays, presented under the three broad themes of *Historical and Cultural Perspectives*; *Images and Heritages*; and *Cultural Urbanism and Planning Strategies* all extend beyond the scope of the communications presented in Barcelona on that occasion.

---

13   The theme of the IPHS 2004 Conference was 'Planning Models and the Culture of Cities'. See the Conference book: *Planning Models and the Culture of Cities. 11th International Planning History Conference* (Barcelona: Centre de Cultura Contemporània de Barcelona, 2004).

PART I
Historial and Cultural Perspectives

Chapter 1

History Builds the Town: On the Uses of History in Twentieth-century City Planning

Michael Hebbert and Wolfgang Sonne

Michael Hebbert is an historian, geographer and town planner. He teaches town planning at the University of Manchester (England), edits the Elsevier research journal *Progress in Planning* and is active with the Urban Design Group and the journal *Municipal Engineer*. He works for the preservation of Manchester's industrial heritage and has published widely on regionalism, urbanism, planning history, and London past and present.

Wolfgang Sonne is lecturer for history and theory of architecture at the Department of Architecture at the University of Strathclyde in Glasgow (Scotland). He studied art history and archaeology in Munich Paris and Berlin and holds a PhD from the Eidgenössische Technische Hochschule in Zurich. He has previously taught at the ETH Zurich, at Harvard University and at the Universität in Vienna. His publications include: *Representing the State. Capital City Planning in the Early Twentieth Century,* Munich, London and New York: Prestel 2003.

Modern urbanism has been profoundly shaped by readings of the past. Town planning is a prospective activity by definition, but its intractable, long-lasting subject-matter forces planners to take a stance towards the history, whether they cling to legacies, memories and precedents or reject them. In the words of Arthur Korn, one of the most outright rejectionists, *History Builds the Town*.[1] In this paper we ask in what sense history built the twentieth century town. What types of historical reference have planners employed? Have they been actively researched and procured, or worn more casually, as loose-fitting myths? Which has counted for more, the particular or the universal?

Awareness of designers' historiographic strategies is well developed in architectural theory.[2] The uses of history in town planning have been less well studied. We argue that they fall into three distinct categories: first of all, history

---

1   Arthur Korn, *History Builds the Town* (London: Lund Humphries, 1953).

2   Wolfgang Sonne, 'Gebaute Geschichtsbilder. Klassizismus, Historismus, Traditionalismus und Modernismus in der Architektur', in Evelyn Schulz and Wolfgang Sonne

has been used as the encyclopaedic source of types and references; second, as the reservoir of collective memory, place-identity and local attachment; and third, as a source of normative justification and purpose for the act of planning. Each type of discourse implies a distinct time pattern and reading of the past. The balance between these modes has shifted in unexpected ways, and each has its risks. A planner with a one-sided sense of history is almost as dangerous as one with none at all.

Lasting Types and Forms

Modern town planning originates at the turn of the twentieth century, at the moment of shock when new technologies began to demonstrate their power to dissolve conventional conceptions of space and time. For those wrestling with this challenge, the first reaction was to look to a timeless formalism.[3] Daniel Hudson Burnham and Edward Herbert Bennett begin their 1909 planning report on Chicago not with statistical data on the modern city but with a general history of classical urban design from Antiquity to the Beaux-Arts.[4] Henry Aldridge opened his *Case for Town Planning. A practical manual for the use of councillors, officers and others engaged in the preparation of town planning schemes* (1915) with a hundred page account of planning as 'one of the oldest of the arts evolved in the slow development of organised civic life in civilised countries'.[5]

In his seminal text for the new discipline, *Town Planning in Practice* (1909), Raymond Unwin emphasised the need for systematic historical research towards a 'classification of the different types of plan which have been evolved in the course of natural growth or have been designed at different periods by human art'.[6] His mentor in this project was the architect Camillo Sitte whose 1889 pamphlet *Der Städtebau nach seinen künstlerischen Grundsätzen* tried to distill lasting rules of urban design from examination of historic layouts. Sitte deliberately used black and white plan analysis in an attempt to abstract from context and derive generic principles for present practice.[7]

It was striking how scholars in the German art-historical tradition brought their discipline's scholarly traditions of periodisation and abstraction of types and forms to bear on the design of cities, largest of human artefacts. Albert Erich Brinckmann

---

(eds.), *Kontinuität und Wandel. Geschichtsbilder in verschiedenen Fächern und Kulturen* (Zurich: vdf Hochschulverlag, 1999), pp. 261–330.

   3   David Harvey, *Condition of Post-Modernity* (Oxford: Blackwell, 1989).

   4   Daniel Hudson Burnham and Edward Herbert Bennett, *Plan of Chicago* (Chicago 1909).

   5   Henry Aldridge, *Case for Town Planning. A practical manual for the use of councillors, officers and others engaged in the preparation of town planning schemes* (London, 1915), p. 9.

   6   Raymond Unwin, *Town Planning in Practice. An Introduction to the Art of Designing Cities and Suburbs*, 2nd edition (London: Allen & Unwin 1911), p. 104.

   7   Camillo Sitte, *Der Städtebau nach seinen künstlerischen Grundsätzen* (Wien: Verlag Carl Graeser, 1889), preface.

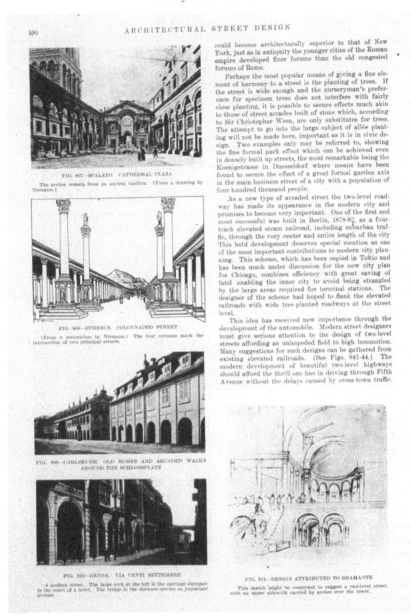

Figure 1.1   Werner Hegemann and Elbert Peets, *The American Vitruvius*, 1922. Arcades and colonnades are arranged in a historic sequence which suggests the eternal recurrence of urban archetypes.

Source: Werner Hegemann and Ebert Peets, *The American Vitruvius. An Architect's Handbook of Civic Art* (New York: Architectural Book Publishing, 1922).

wrote both on the design of squares (1908) and on town planning in general (1920).[8] Cornelius Gurlitt's copious historical writing was matched by a manual of urban design in 1920.[9] The most ambitious compendium combined Germanic formalism with American pragmatism. Werner Hegemann's and Elbert Peets' *American Vitruvius: An Architects' Handbook of Civic Art* (1922) drew examples from 3,000 years of urban history with more than 1,200 illustrations.[10] Historic variety was ordered according to universal urban elements, creating subtle thematic connections across long periods.

This way of constructing the affiliation of motifs resembled Aby Warburg's contemporanous attempts to classify the painted antique *pathos formula* in the large tables of his 'Mnemosyne-Atlas'.[11] Hegemann and Peets were not writing an account of real historic influences or trying to conjure a return of archaic forms, but offering contemporary designers an encyclopedic scheme of urban elements, with a clear preference for a timeless classicism of axial layout and rhythmic enclosure.[12]

This claim for the universality of Civic Art was compromised once Le Corbusier and the Charter of Athens had explicitly jettisoned all historic types and forms in favour of a radically new and puritan urbanism, based upon the functional logic of the modern city. But it soon became apparent to town planners that their clients expected them to inject some humanism into the austere doctrine of functional separation set out in the Charter of Athens. The issue of human association dominated postwar meetings of CIAM, caused the breakaway of younger modernists in Team X in 1953, and triggered the emergence of the subdiscipline of urban design in 1956. The search for the heart of the city, its symbolic and interactive essence, brought modernism back to the interrogation of history.[13]

---

8   Albert Erich Brinckmann, *Platz und Monument* (Berlin, 1908); Albert Erich Brinckmann, *Deutsche Stadtbaukunst in der Vergangenheit* (Frankfurt am Main, 1911); Albert Erich Brinckmann, *Stadtbaukunst. Geschichtliche Querschnitte und neuzeitliche Ziele* (Berlin, 1920).

9   Cornelius Gurlitt, *Historische Stadtbilder*, 12 vols. (Berlin, 1901–12); Cornelius Gurlitt, *Handbuch des Städtebaues* (Berlin, 1920).

10   Werner Hegemann and Ebert Peets, *The American Vitruvius. An Architect's Handbook of Civic Art* (New York: Architectural Book Publishing, 1922). Cf. Christiane Craseman Collins, *Werner Hegemann and the Search for Universal Urbanism* (New York: Norton, 2005).

11   Martin Warnke (ed.), *Aby Warburg. Der Bilderatlas Mnemosyne* (Berlin, 2000).

12   Wolfgang Sonne, 'Bilder, Geschichte und Architektur. Drei wesentliche Bestandteile der Städtebautheorie in Werner Hegemanns und Elbert Peets' American Vitruvius', in *Scholion. Bulletin der Stiftung Bibliothek Werner Oechslin*, 2 (2002): pp. 122–133; Wolfgang Sonne, 'Images, History, and Architecture. Three Crucial Elements of an Appropriate Urban Design Theory in Hegemann and Peets' "American Vitruvius"', in *Proceedings of the Conference 'The New Civic Art, Werner Hegemann and International Exchanges'* (Miami, 2005), in print.

13   Sigfried Giedion, Jose Luis Sert and Fernand Léger, 'Neun Punkte über: Monumentalität – ein menschliches Bedürfnis' (1943), in Sigfried Giedion, *Architektur und Gemeinschaft. Tagebuch einer Entwicklung* (Hamburg: Rowohlt, 1956), pp. 40–42; and Jacqueline Tyrwhitt,

PLATE 1

A: A CHILD'S IMAGE
OF A SQUARE
DRAWING BY A NINE-
YEAR-OLD GIRL

From Hans Friedrich Geist,
"Paul Klee und die Welt des
Kindes," *Werk* (June, 1950)

B: PARIS, PLACE ROYALE
(PLACE DES VOSGES)

COPPER ENGRAVING

From Martin Zeiller, *Topogra-
phia Galliae* . . .

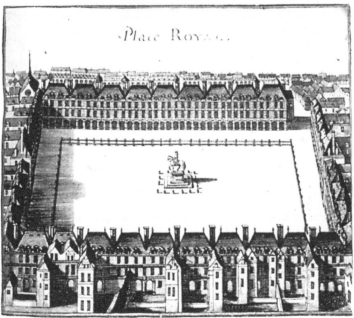

Figure 1.2   Paul Zucker, The archetype of a square in a children's drawing and the Place Henri IV in Paris, in *Town and Square*, 1959. The eternal type of an urban element is represented by the 'innocent hand' of a child.

Source: Paul Zucker, Town and Square. From the Agora to the Village Green (New York and London: Columbia University Press 1959).

Kevin Lynch and Edmund Bacon defined urban design in terms of the analysis of historic streets and quarters, reestablishing the seminal importance of a Grand Tour of Italian exemplars.[14] Joseph Rykwert went further back, looking to ancient Rome for source of symbolic meaning in urban design.[15] At MIT, Stanford Anderson initated a major study into the history of streets and their dual generic role as a basis of circulation and a setting for exchange.[16] And, in an early and ambitious contribution, Paul Zucker studied the history of the urban square, showing it to be a psychologically determined archetype of universal relevance at every stage of development and every scale of settlement. In his opening pair of illustrations a child's drawing reveals the *Ur-form* of a square which also informed in Henri IV's Place Royale in Paris.

The 'Paradox of History', for Zucker, was that types and forms could remain constant however radically functions change over the epochs of urban history.[17]

Towards the close of the century the broad current of postmodern urbanism became increasingly receptive to historical material that could inform revivalism. There were synoptic narratives such as the huge pictorial atlas of Leonardo Benevolo's *Storia della città* and the encyclopedic multivolume *Storia dell'urbanistica*,[18] and a revival in art-historical contributions such as Wolfgang Braunfels' *Abendländische Stadtbaukunst* and Michael Hesse's overview of *Stadtarchitektur*.[19] Spiro Kostof, another architectural historian, contributed the most subtle and wide-ranging study of historic types and form in his final pair of books *The City Shaped* and *The City Assembled*.[20] Like in the *American Vitruvius* from the 1920s, the structure was thematic, not chronological, deliberately inviting the modern practitioner to reflect and translate. In these years practising urbanists contributed actively to historical investigation. Rob Krier's teaching and practice in Stuttgart gave rise to

---

Jose Luis Sert, Ernesto Nathan Rogers, *The Heart of the City. Towards the Humanisation of Urban Life* (New York: Pellegrini and Cudahy, 1952).

14 Kevin Lynch, *The Image of the City* (Cambridge, Mass.: MIT Press, 1960); and Edmund Bacon, *Design of Cities* (London: Thames & Hudson, 1967).

15 Joseph Rykwert, *The Idea of a Town. The Anthropology of Urban Form in Rome, Italy and the Ancient World* (Cambridge, Mass.: MIT Press, 1988).

16 Stanford Anderson (ed.), *On Streets* (Cambridge, Mass.: MIT Press, 1978).

17 Paul Zucker, *Town and Square. From the Agora to the Village Green* (New York and London: Columbia University Press, 1959), p. 18.

18 Leonardo Benevolo, *Storia della città* (Rome and Bari: Laterza, 1975); *Storia dell'urbanistica*, 12 vols. (Rome and Bari: Laterza, 1976–91).

19 Wolfgang Braunfels, *Abendländische Stadtbaukunst. Herrschaftform und Baugestalt* (Cologne: Du Mont, 1976); Michael Hesse, *Stadtarchitektur. Fallbeispiele von der Antike bis zur Gegenwart* (Cologne: Deubner Verlag, 2003).

20 Spiro Kostof, *The City Shaped. Urban Patterns and Meanings through History* (London: Thames & Hudson, 1991); and Spiro Kostof, *The City Assembled. The Elements of Urban Form through History* (London: Thames & Hudson, 1992).

his study of generic types of street and block pattern,[21] Phillipe Panerai and fellow members of the Versailles school produced their seminal analysis of the transition from nineteenth to twentieth century urban form,[22] and Allan Jacobs contributed meticulously researched typologies of urban thoroughfares.[23]

For the neo-traditionalist wing the revival of historic types and forms became a matter of doctrine. Advocacy of street and square got entangled in arguments for masonry and timber building, vertical windows, and adherence to norms of classical detailing. Leon Krier was a polemical advocate of eternal principles of urban design, to which all good cities must conform.[24] The wit and elegance of his work made up for the narrowness of its frame of reference. A more synoptic attempt to distil a systematic theory of urbanism from past types was the *New Civic Art. Elements of Town Planning* compiled by Andres Duany, Elizabeth Plater-Zyberk and Robert Alminana in 2003. Title and organisation pay deliberate homage to Hegemann's and Peets' *American Vitruvius. An Architect's Handbook of Civic Art* (1922). In terms of its display as well as its contents the *New Civic Art* is anything but new – following established models, it draws from all historical periods and sources, including Hegemann's book and modernist planning projects. Its concept of history is a pragmatic traditionalism: every historic solution may be chosen if it appears to be good. No barriers are constructed between past and present, for as the authors say in the preface: 'this collection is only a manual of proven practice'.[25]

New Urbanism's strategy – codify history and use the codes as a basis for present action – has been heavily criticised for its formulaic manipulation of appearances. Its most disturbing repercussion has been to divorce town planning from the creative world of architecture. But doctrinal neo-traditonalism is a fringe movement, unrepresentative of the generality of planning practice for which knowledge of generic types and forms of city-building is part of the essential process of learning the vocabulary of urbanism.

---

21 Rob Krier, *Stadtraum in Theorie und Praxis* (Stuttgart: Krämer, 1975); engl.: *Urban Space* (London: Academy Editions, 1979).

22 Philippe Panerai, Jean Castex, Jean-Charles Depaule, *Formes Urbaines. De l'ilot a la barre* (Paris : Dunod, 1977); engl.: *Urban Forms. Death and Life of the Urban Block* (Oxford: Architectural Press, 2004).

23 Allan B. Jacobs, *Great Streets* (New York: Random House, 1993); Allan B. Jacobs, Elizabeth Macdonald, Yodan Rofé, *The Boulevard Book: history, evolution, design of multiway boulevards* (Cambridge, Mass.: MIT Press, 2002). Cf. N.R. Fyfe (ed.), *Images of the Street* (London: Routledge, 1998).

24 Leon Krier, *Architecture. Choice or Fate*, Windsor: Papadakis 1998.

25 Andres Duany, Elisabeth Plater-Zyberk, Robert Alminana, *The New Civic Art. Elements of Town Planning* (New Yrok: Rizzoli, 2003), p. 9. Cf. Michael Hebbert, 'New Urbanism – the movement in context', in *Built Environment*, 29, 3, (2003): pp. 193–210.

**Figure 1.3   Patrick Geddes, Arbor Saeculorum – the tree of the centuries, 1895.
The process of history is represented as organic growth.**
Source: Volker Welter, Biopolis. Patrick Geddes and the City of Life (Cambridge, Mass.:
MIT Press, 2002).

History as Local Memory

Town planning may draw on universal types but it is also an encounter with specific place and people. Local distinctiveness arises from a specific history which the planner needs to appreciate. Perhaps the most distinctive theory of this kind informed the work of Scottish town planning activist and theoretician Patrick Geddes. His view of history as a continuous process of growth was clearly marked by biological overtones and may be best represented by his design for a stained glass window, showing the 'Arbor Saeculorum', the tree of the centuries, in 1892.

The different historic cultures unfold like a growing tree, each oscillating between the 'temporal' and the 'spiritual powers'.[26] Contemporary time is the quasi-biological result of all that has gone before.

'A city is more than a place in space', Geddes wrote, 'it is a drama in time' proceeding in phases that he likened to 'the layers of a coral reef in which each generation constructs its characteristic stony skeleton as a contribution to the growing yet dying and wearying whole'.[27] Geddes arranged his famous travelling Cities and Town Planning Exhibition – starting in Chelsea in 1911 and destroyed in the Pacific Sea in 1914 – according to a chronological order beginning with the 'origin and rise of cities' and finishing with the latest developments of the Garden Cities.[28] Contemporary planning emerges directly from history, not only firmly rooted in time but also in space. Therefore Geddes's most widely-communicated idea was the need for intensive, historically-based local survey as a basis for action. Only through intense process of sharing collective memory could the future be projected.[29] The Outlook Tower embodied his approach, being both a museum of civic development and a centre for urbanism, with a rooftop *camera obscura* looking out on contemporary Edinburgh.[30] While Geddes brought his own peculiar methodology, others took different routes to the same end of an intensive, respectful survey. In France the archival and cartographic tradition of Paris historian Marcel

26 Volker Welter, *Biopolis. Patrick Geddes and the City of Life* (Cambridge, Mass.: MIT Press, 2002), pp. 88–90.

27 Helen Meller (ed.), *The Ideal City* (Leicester: Leicester University Press, 1979), pp. 79, 82.

28 Volker Welter, *Biopolis. Patrick Geddes and the City of Life* (Cambridge, Mass.: MIT Press, 2002), pp. 124–127.

29 Patrick Geddes, *Cities in Evolution. An Introduction to the Town Planning Movement and to the Study of Civics* (London: Williams and Norgate, 1915), pp. 329–375.

30 Helen Meller, *Patrick Geddes. Social Evolutionist and City Planner* (London: Routledge, 1990).

Poëte,[31] in Germany an emphasis on regional building types and land ownership patterns, based on the field discipline of geographical morphology.[32]

In the mid-century, modernist town planning gave a completely different meaning to the Geddesian concept of survey before plan. The survey became a fact-collecting exercise, intended to define an empirical basis for technical action. But even in this period historical contributions like Frederick Hiorns *Town-Building in History* (1956) could display a local and evolutionary approach. A London County Council architect of the old school, he saw urban design in terms of continuity and locality:

> What we know as tradition is (and always was) but a natural and continuous flow of development, a condition of fluidity that adapts itself to the changing circumstances of time, merging the past into the present, and the present into the future. The idea that creative work becomes 'contemporary' only when torn from any recognisable association with the past is entirely fallacious. Regarded in this way, the course of building in the present century has produced excellent and varied examples of modern design – 'Georgian' in quality – free of the taint of freakish, exotic importations. In the continuation and logical development of this lies our hope for the future.[33]

In the 1960s and 1970s there was a broad rediscovery of the local dimension. In the Anglo-Saxon world the main impetus was the conservation movement and desire to protect the character of historic settlements and monuments. At the seminal conference on *The Historian and the City*, organised by the Massachussetts Institute of Technology and Harvard University in 1962 Christopher Tunnard of Yale called for city planning to base itself more soundly on more scholarly research into what he called the historical 'build' of the cities – their blocks and urban spaces.[34]

This was exactly what was beginning to happen in continental Europe as Italian urbanists developed their techniques of detailed mapping of historic types. A first

---

31 Préfecture du Département de la Seine, Commission d'Extension de Paris, *Aperçu historique* (Paris, 1913); Marcel Poëte, *Une vie de cité. Paris de sa naissance à nos jours*, 4 vols. (Paris, 1924–31); Marcel Poëte, *Introduction à l'urbanisme. L'évolution des villes* (Paris, 1929). Cf. Donatella Calabi, 'Marcel Poëte. Pioneer of 'l'urbanisme' and Defender of 'l'histoire des villes'', in *Planning Perspectives*, 11/4 (1996): pp. 413–436; and Donatella Calabi, *Parigi anni venti. Marcel Poëte e le origini della storia urbana* (Venice, 1997).

32 Michael R.G. Conzen, *The Urban Landscape. Historical development and management*, Institute of British Geographers Special Publication No. 13, ed. by Jeremy Whitehand (London: Academic Press, 1981).

33 Frederick R. Hiorns, *Town-Building in History. An Outline Review of Conditions, Influences, Ideas, and Methods Affecting 'Planned' Towns through Five Thousand Years* (London etc: George G. Harrap & Co, 1956), p. 412. Cf. Peter Larkham, 'The place of conservation in the UK reconstruction plans of 1942–52', in *Planning Perspectives*, 18 (2003): pp. 295–324.

34 Christopher Tunnard, 'The Customary and the Characteristic. A Note on the Pursuit of City Planning History', in Oscar Handlin and John Burchard (eds.), *The Historian and the City* (Cambridge : MIT Press, 1963), pp. 216–224; cf. also pp. 266–267.

step had been the typological plans for a *Storia operante* by Saverio Muratori in the late 1950s.[35] Aldo Rossi's *Architettura della Città* (1966) then lifted typomorphology from a specialised scholarly pursuit into a general basis for urban design.[36] Like Colin Rowe and Fred Koetter's *Collage City* in the Anglo-Saxon world, it created a new awareness of inherited architectural form as a primary factor of a city's identity and character – a total reversal of the functionalist ideology in which urban space was a mere residual.[37] The most important historical method used was the simple, perhaps superficial, technique of studying building and urban space patterns through figure-ground analysis.[38] Combined with documentary and oral history, and public participation, it provides a basis for town planning in the new century to demonstrate a clear pedigree in collective memory. We expect urbanism to have some foundations in local character and distinctiveness. Preservation of local memory is one element in the equation. If it were the only concern, urbanism would be parochial, narcissistic and backward-looking, but if respect for collective memory is entirely absent, planning betrays the past. This is the crux of urbanism in any city, but especially where memories are painful, such as Berlin.[39]

The Big Canvas

There is a third historiographic strand in modern city planning. Not the excavation of a particular and local characteristics, nor the search for universal types and principles: but a larger narrative history, interpreting the past, construing the future, and defining the actor's place within it. In the history of twentieth century ideas the planning movement was closely associated with what Karl Popper called historicism, the belief that reformers should both diagnose impending change and play the part of a social midwife to bring it about.[40] One of the more dangerously fascinating time-pattern was life-cycle of birth maturity and death of cities and civilisations: Plato applied it prophetically to the destiny of Greek cities in his *Republic*, and it recurred over the years in Machiavelli, Vico, Spengler and Toynbee.[41]

In the mid twentieth century the prevailing ideology of avantgardism made town planners indifferent to most types of history. But the myth of death and rebirth seemed to speak for town planning. As in Marcel Lods's cartoon-like images of his scheme for postwar Mainz in 1946, it pitched a dark and narrow image of the city of the past

---

35 Saverio Muratori, *Studi per una operante storia urbana di Venezia* (Rome, 1959).

36 Aldo Rossi, *L'Architettura della città* (Padua: Marsilio, 1966).

37 Colin Rowe and Fred Koetter, *Collage City* (Cambridge, Mass.: MIT Press, 1978).

38 Anne Vernez Moudon, 'Urban morphology as an emerging interdisciplinary field', in *Urban Morphology*, 1 (1997): pp. 3–10.

39 Michael Hebbert, 'The Street as Locus of Collective Memory', *Environment and Planning D: Society and Space,* 23/4 (2005): 581–596.

40 Karl Popper, *The Poverty of Historicism* (London: Routledge Kegan Paul, 1957), p. 49.

41 Ibid., p. 110.

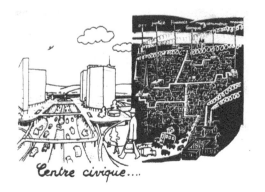

Figure 1.4   Marcel Lods, Reconstruction plans for Mainz, 1946. The contrast between a gloomy past and a bright future is shown in an explicitly clear comic strip.

Source: Vittorio Magnago Lampugnani and Romana Schneider (eds.), Moderne Architektur in Deutschland 1900 bis 1950. Expressionismus und Neue Sachlichkeit (Stuttgart: Hatje, 1994).

against bright and wide panoramas of a future reconstruction. In the mid-century of high modernism, the less attention was paid to the value of historic examples or to collective memory, the greater the fascination with big-canvas narratives. In Lewis Mumford's *Culture of Cities* (1938), urban history is set out panoramically as a cycle of growth and decay.[42] The narrative rhetorically conveys (in the words of Sidney Checkland) 'a sense of ghastly, relentlessly cumulative error leading to social corruption and impotence'.[43] But the pessimism is offset by a promise of resurrection through city planning and its parkways, neighbourhood units, planned decentralisation and garden-cities.

Some mid-century historians documented the past from a simple conviction that it was passing. Ernst Egli's *Geschichte des Städtebaus*, published in three volumes 1959–67, was the largest attempt to write a general history of urban design at his time. The author was a practising modern architect, especially in Austria and Turkey, before he became professor for town planning in Zurich. His *History of Urban Design* presents the city in history as something entirely apart from the present city: 'One thing is for sure, the era of the old city is finishing.'[44] He described his entire project as just an unsentimental glance back before turning towards a different future.[45]

A similar attitude had been taken by Erwin Gutkind. His *International History of City Development* – published in eight volumes and more than 4,000 pages between 1964 and 1972 – was the largest celebration of the historical diversity of towns and town planning. But the author believed he was describing an irrevocably lost world. There could be no survival for the conventional city.[46] Like Mumford he believed that dense urban populations must be dispersed, and nature allowed to grow where streets and buildings once stood: 'The original conception of a city – a conception that has lasted for 5000 years with only minor modifications – is now approaching its end.'[47] As history was past, the future had to be new: 'It is our task, at once inspiring and terrifying, to begin a new chapter in the history of human settlement.'[48]

The paradoxical trope reappears again and again, and the more radical the future vision, the deeper its footings in an argument of historical necessity. Giedion's hugely influential *Space Time and Architecture. The growth of a new tradition* (1941) opens with a philosophy of history, which he sees as a dynamic process of discovering your destiny and enlarging your ambitions: 'This does not mean that we should

---

42 Lewis Mumford, *The Culture of Cities* (New York: Harcourt, Brace & Co 1938), p. 283.

43 Harold James Dyos (ed.), *The Study of Urban History* (New York: St. Martin's Press, 1968), p. 346.

44 Ernst Egli, *Geschichte des Städtebaus*, vol. 1, *Die alte Welt* (Zürich and Stuttgart: Eugen Rentsch 1959), p. 9.

45 Ibid., p. 9.

46 Erwin Anton Gutkind, *The Twilight of Cities* (New York and London: The Free Press of Glencoe, 1962), p. 56.

47 Erwin Anton Gutkind, *International History of City Development*, vol. 1, *Urban Development in Central Europe* (London: Collier-Macmillan 1964), pp. 5–6.

48 Ibid., p. 6.

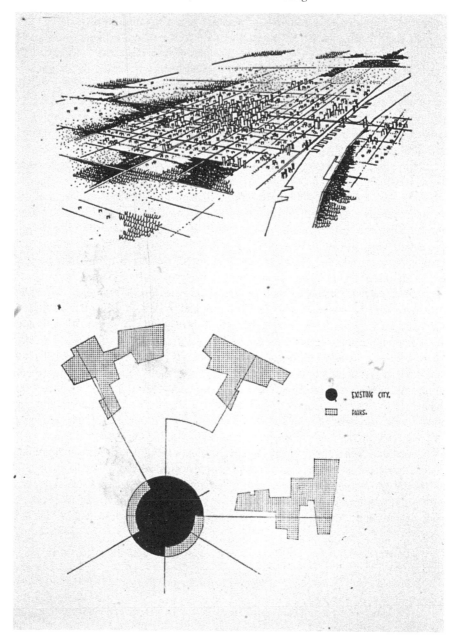

**Figure 1.5   Erwin Anton Gutkind, The existing city, in *The Twilight of Cities*,
1962. The traditional city follows the pattern of a dense urban
fabric surrounded by countryside.**

Source: Erwin Anton Gutkind, *The Twilight of Cities* (New York and London: The Free Press
of Glencoe, 1962).

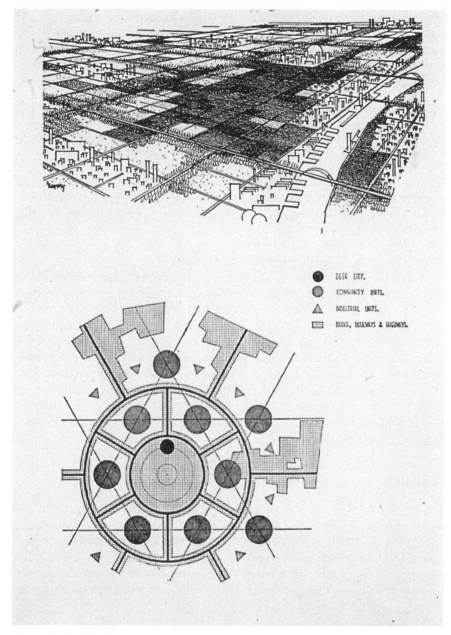

Figure 1.6 **Erwin Anton Gutkind, The future city, in *The Twilight of Cities*, 1962. The planner envisions the total destruction of the historic city and the creation of a new network of settlement.**

Source: Erwin Anton Gutkind, *The Twilight of Cities* (New York and London: The Free Press of Glencoe, 1962).

copy the forms and attitudes of bygone periods, as the nineteenth century did, but that we should conduct our lives against a much wider historical background'.[49] The implication for twentieth century urbanists was that they should be brave enough to abolish everything familiar in the urban setting, from the geometry of streets to the large-scale landscape-pattern of town and country:

> The use of a new and larger scale in town planning which would coincide with the scale already being used in the parkway system is an imperative necessity for the salvation of the city. This scale must permeate all urban projects. It is closely connected with the space-time conception of our period.[50]

Eliel Saarinen's *The City. Its Growth – Its Decay – Its Future* (1943) spelled out a similar hypothesis.[51] Part One of the book – 'The Past' – described the organic growth of cities since the dawn of civilisation and their twentieth century affliction by 'urban disease'. Part Two – 'Toward the Future' – prescribed the remedy, comprehensive cleansing and reorganisation of cities around modern principles of traffic organisation, separated from the developments of the past.

And so we came at last to our title piece, Arthur Korn's *History Builds the Town* (1953). Korn had been the communist spokesman for the 'proletarian' tendency in CIAM IV, and was the driving force behind the radical MARS plan for London.[52] Disappointed by what he felt to be the tame and compromising reconstruction of postwar London, his statement of the 'correct theory' of town planning was a particularly ruthless variant of clean-sweep doctrine. The history that should build the town was a dialectical Marxism. Three quarters of the book describes the historic processes of town-making. But in the last quarter the slate is wiped clean so modern welfare-based principles can be applied in an urbanism of 'Theory and Practice'. History builds the town in the sense that it confirms the destiny of the town planner, as agent of transformation.

Conclusion

Town planners have never been more in need of historical appreciation and imagination. Nimbyism, urban sprawl, the motorcar versus the pedestrian, design codes, outdoor advertising, tall buildings controls, blob architecture: all the toughest contemporary controversies involve some judgement of the claims of the past. But what sort of history do planners need to know? Looking back on the corpus of planning literature we have distinguished three types. One has to do with a general knowledge

---

49 Sigfried Giedion, Space, *Time and Architecture. The Growth of a New Tradition* (Cambridge, Mass.: Harvard University Press, 1941), p. 8.

50 Ibid., p. 569.

51 Eliel Saarinen, *The City. Its Growth – Its Decay – Its Future* (New York: Reinhold, 1943).

52 John Robert Gold, *The Experience of Modernism. Modern Architects and the Future City 1928–53* (London: E & FN Spon, 1997).

of what cities are and an ability to recognise lasting components and attributes. The second has to do with recognising individuality and coaxing memory. And the third is the great mirror of history. In the twentieth century the three historiographies had an inverse relation. The more planners became fascinated by the image of their own role as agents of historical change, the less attention they paid to the study of generic types and local idiosyncrasies. As the power of the mirror waned, other types of historical sensibility revived.

The perspective is different in the twenty-first century. For the first time in 50 years urbanists are seeing themselves again in a world-historical perspective. This time the mirror shows them not as agents of modernity and technical progress but in a role of environmental stewardship. Climate change has restored the big time frame and made local actors aware of global consequences. It has reinvigorated urbanism, because carbon reduction strategies demand a curbing of urban sprawl and a search for compact, energy-efficient human settlements. This in turn has produced a new appreciation of traditional typologies and regional variances. Looking at the trend of their times, Siegfried Giedion, Arthur Korn and Erwin Gutkind concluded that all knowledge of the human settlements through the centuries had become redundant, a mere antiquarian curiosity. It would be impossible to reach that conclusion today.

Chapter 2

The Cultural Dimension of Urban Planning Strategies: An Historical Perspective

Robert Freestone and Chris Gibson

Robert Freestone is Associate Professor of Planning and Urban Development at the University of New South Wales, Sydney (Australia). His main research interests are in the evolution of urban planning theory and practice, metropolitan restructuring, and the built landscape. He is the author of *Model Communities: The garden city movement in Australia* (1989) *and editor of Urban Planning in a Changing World* (2000).

Chris Gibson is Senior Lecturer in Geography in the School of Earth and Environmental Sciences, University of Wollongong, New South Wales (Australia). He has published widely on the creative industries, urban development and popular culture, particularly in the Australian context. His books include *Sound Tracks: Popular Music, Identity and Place* (2003) *and Deadly Sounds, Deadly Places: Contemporary Aboriginal Music in Australia* (2004).

Cultural activities are of growing significance to urban and regional economies. The term 'cultural economy' defines this intersection across a broad spectrum of creative endeavors including music, film, television, drama, art, design, and media. These activities have assumed importance in urban policy and city planning singly or collectively as key sectors for often interconnected initiatives in economic development, urban regeneration, place-making, urban design, and social planning. The 'cultural turn' in city planning looms as an important dimension to be factored into considerations of future urban change.

This chapter develops an orientation to the cultural–planning nexus by constructing an interpretation of past trends segueing to current initiatives employing a planning history perspective. The overall aim is to offer a broad, heuristic reading of an evolving, strengthening but not always unproblematical interplay between cultural activity and urban planning as a vital consideration for the urban prospect. The chapter employs the generalised chronological framework of a succession of historical cultural planning paradigms from the turn of the twentieth century onward.

Culture and Planning

'Culture' is arguably one of the richest and most complex concepts in any language.[1] There are hundreds of definitions and a myriad of expressions in many locations. Culture is material and symbolic, belonging and being, pattern and process, macro and micro, corporate and public, product and public good. 'Culture', it has been said, 'is everything'.[2]

Sharon Zukin's *The Cultures of Cities* – revealingly titled without even a passing reference to Lewis Mumford's 1938 classic *The Culture of Cities* – conveys well the hybridity and ambiguity of contemporary urban culture.[3] What was once an elite 'fait accompli' has exploded with spectacular secularisation into an increasingly flexible and contested 'agent of change'. Culture has become a business of cities, the significance of this cultural economy now measured in terms of employment, expenditure, wages, and multipliers. Places themselves – their architecture, their townscapes, their street life – have become sources of inspiration in the creative process. Culture is also about controlling cities – their images, meanings, comparative advantage, destinies, and corporate appeal. There are rich pickings here. Zukin herself heads off into discussions of fear, security, and social exclusion. For other observers the gaze comes to rest on the cultural industries as one of the fastest growing industrial sectors in the world.[4] David Harvey extends the argument into an interlinking of culture, capital and commodification to bemoan city degenerating 'into compliance with the prevailing order'.[5]

This burgeoning of perceptions and realities of cultural activity lies at the heart of its relationship with modern urban planning. The current cultural turn may be explosive, but it is not sudden; indeed, the interlocking of cultural institutions and urban development is time-honored. The historical narrative outlined here describes a long, slow courtship based largely on the expression of patrician values democratizing to infect planning on many more fronts, changing its nature and shaping its direction as well as revealing its shortcomings and causing new dilemmas. As culture has become more braided and contested, so too has planning. The unfolding interaction of these two multivalent concepts, and with one (culture) ultimately subsuming the other (planning), describes an extraordinary and unfurling complexity of possibilities:

---

1   R. Williams, *Culture and Society 1780–1950*, (Harmondsworth: Penguin, 1966).

2   D. Mitchell, D., 'There's no such thing as culture: towards a reconceptualization of the idea of culture in geography', *Transactions of the Institute of British Geographers*, 20 (1995): pp. 102–116.

3   S. Zukin, *The Cultures of Cities* (Cambridge: Blackwell, 1995).

4   A.J. Scott, *The Cultural Economy of Cities* (Los Angeles: Sage, 2000); and A.J. Scott, 'Cultural-Products Industries and Urban Economic Development: Prospects for Growth and Market Contestation in Global Context', *Urban Affairs Review*, 39 (2004): pp. 461–490.

5   D. Harvey, *Spaces of Hope* (Edinburgh: Edinburgh University Press, 2000), p. 169.

Table 2.1   An evolution of physical-economic-cultural planning paradigms

Period	Paradigms	Theorists and practitioners	Places, Plans and Exemplars
1900s–1910s	City as a work of art	Daniel Burnham	The models of Paris and Vienna; city beautiful movement; Plan of Chicago; Plan for Canberra
1910s–1950s	Cultural zonation	Harland Bartholomew Patrick Abercrombie	Civic-cultural centers; neighborhood civic facilities; city functional and post WW2 master plans
1960s–1970s	Flagship facilities	Robert Moses	Lincoln Centre; JFK Centre; Sydney Opera House
1960s–1970s	Cultures of communities	Jane Jacobs	Community arts facilities; heritage movement; community cultural development; social planning
1980s–1990s	Culture in urban development	Progressive city administrations Pasqual Maragall Sharon Zukin	Cultural regeneration and cultural industries strategies; festival marketplaces; local economic development; European Capital of Culture; Barcelona; Bilbao; Baltimore; Glasgow; Manchester
1990s–2000s	The creative city	Charles Landry Richard Florida Allen Scott	Arts and cultural planning strategies; urban design; cultural precincts; cultural tourism; Huddersfield; Helsinki; Berlin

Land-use and culture are fundamental natural and human phenomena, but the combined notion and practice of culture and planning conjure up a tension between not only tradition, resistance and change; heritage and contemporary cultural expression, but also the ideals of cultural rights, equity and amenity.[6]

A Planning History Perspective

The accommodation of cultural activities through the rise of modern city planning has been acknowledged. The interpretation reported here has benefited from the

---

6   G. Evans, *Cultural Planning: An urban renaissance* (London: Routledge, 2001), p. 1.

historical surveys of Bassett, Comedia, Evans, Hall and Young.[7] The distinctiveness of our approach is to link culture more tightly into the history of mainstream urban planning theory and practice, looking more intensively than previously at the formative decades. As impossibly expansive as the treatment is, several caveats are in order. The coverage is basically confined to big cities in the twentieth century. The primary focus is the shaping of the built environment by spatial policies, the so-called 'physical aspects of public culture'.[8] An Anglo-American perspective predominates. And a major theme is the tension and rapprochement between elite and popular culture in spatial terms. The conventional paradigmatic/chronological framework has both power in retelling history and weaknesses in dealing with departures and temporal outriders. Our six major phases are offered only as a first orientation to appreciating a rich nexus between cultural economy and city planning in evolution (Table 2.1). They represent not so much a singular inexorable trajectory as capture the *zeitgeist* of different eras with some or all approaches still represented today in different places.

The City as a Work of Art

In the early nineteenth century, urban cultural participation is often depicted as a democratic experience, with Shakespeare and acrobats jostling in the playhouse, and Bach and popular favourites airing in the concert hall. At the approach of a new century, high art and popular culture became more institutionally segregated.[9] This division was reflected in urban governance. High culture would become of most interest to grand manner planning as a direct reflection of desired societal values while the low culture of popular taste remained embedded in commercial activity and so of interest only in a regulatory sense through building, fire and public nuisance codes.

Haussmann's Paris expresses this bifurcation in spatial terms. Two levels of urban experience were in evidence: the big picture and the fine grain; top down and bottom up. The grand boulevards showcasing citadels of high art such as L'Opera (1861) were the 'theatrical spaces, designed for display' while the 'ordinary streets behind them [were] quite different: places of everyday life'.[10] However,

---

7   K. Bassett, 'Urban cultural strategies and urban regeneration: a case study and critique', *Environment and Planning A*, 25, (1993): pp. 1773–1788; Comedia, *Glasgow: The creative city and its cultural economy* (Gloucester: Comedia, 1991); Evans, *Cultural Planning: An urban renaissance*; P. Hall, *Cities of Tomorrow: An Intellectual History of Urban Planning and Design in the Twentieth Century*, 3rd ed. (Oxford: Blackwell, 2002); and G. Young, 'The Cultural Reinvention of Planning', PhD thesis (University of New South Wales, 2005).

8   Evans, *Cultural Planning: An urban renaissance*, p.2.

9   I.W. Levine, *Highbrow/lowbrow: The emergence of cultural hierarchy in America* (Cambridge: Harvard University Press, 1988).

10 P. Hall, *Cities in Civilisation: Culture, Innovation and Urban Order* (London Weidenfeld and Nicholson, 1998), p. 721.

the whole restructuring process of haussmanisation unleashed powerful forces of embourgeoisement in Paris and other European capitals where the same high : low cultural division was codified in the urban landscape. Some cities established agglomerations of elite cultural institutions, such as the museum quarters of South Kensington in London, Amsterdam's 'Museumplein', and Berlin 'Museumminsel'. Vienna razed its ancient fortifications and replaced them with parkland and cultural institutions. In the new world, the emergent bi-level planning was captured in the 1893 Chicago Worlds Fair with its schism between the high culture of the pavilion city and the lowbrow entertainment of the Midway Plaisance.

Early twentieth century planning was dominated by 'high culture' defined in bourgeois terms as a moral force for beautification and improvement. This revolved around museums, libraries, public gardens, art galleries and concert halls. A defining characteristic of these facilities was to display of wealth rather than to generate it directly. It was a long way from the dancehalls and pubs of everyday culture, and this separation of elite from mass culture was similarly expressed in spatial terms.

Ebenezer Howard's[11] (1898) famous diagram of Garden City was an ideal expression of a new form of post-industrial revolution urban settlement. While the main employment generators of cottage-industries hugged the circular periphery, at the centre were grouped 'larger public buildings ... each standing in its own ample grounds'. These included not only the obligatory town hall plus a hospital, but a predominance of cultural venues: the 'principal concert and lecture hall', a library, theatre, museum and 'picture-gallery'. They surrounded a large central garden and were separated from the town proper by an encircling 'central park' of 145 acres.

Translation into practice came early as welfare capitalists took up the notion of places self-contained for cultural facilities in planned communities. At the wider city scale, the development of new cultural institutions became part of the process of differentiating high art from mass entertainment. Construction of imposing buildings stood in stark contrast to the music halls, taverns, and circuses in commercial and residential districts (Figure 2.1). City beautiful planners developed more comprehensive schemes for elite cultural precincts.

The classic and statement of a city beautiful in cultural planning terms is Daniel Burnham and Edward Bennett's *Plan of Chicago* (1909). While not ignoring a new suburban culture birthed in park fieldhouses around 'adult citizenship' classes, its main ambition was creation of an American Paris downtown. Burnham recognised 'art as a source of wealth and moral influence', starting with the many students whose training in quality facilities of sufficient scale would help raise the 'standard of public taste'. To complement the civic centre devoted to 'the administrative point of view' would be a similar monumental group plan devoted to the 'intellectual life' of the city. Sited in Grant Park near the waterfront and clustered around the Field Museum, the Art Institute and a new public library, this acropolis of culture would showcase art, education, edification, taste, philanthropy, and civic-mindedness.

---

11 E. Howard, *To–Morrow: A Peaceful Path to Real Reform* (1898), annotated edition edited by P. Hall, D. Hardy and C. Ward (London: Routledge, 2003).

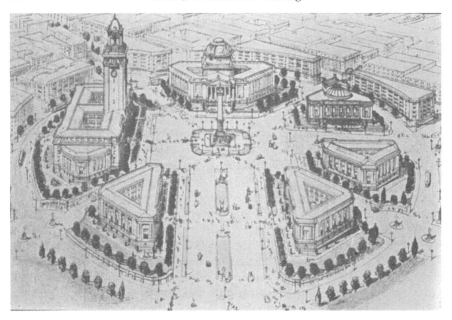

Figure 2.1   **Early association of administrative and cultural functions in a city beautiful complex: Virgil Bogue's civic centre in his Plan of Seattle (1911)**

Source: John Sulman, *An Introduction to the Study of Town Planning in Australia* (Sydney: Government Printer, 1921).

Cultural Zonation

The idea of the formal cultural precinct is one of the most enduring in modernist planning, a classic expression of its penchant for mono-functional zoning. So entrenched was this segregationist thinking that even the preceding view of the whole city as a work of art was lost as the arts and urban planning diverged. The cultural aspirations of citizens would be narrowly confined to particular urban spaces. The methodology of the civic survey holding sway into the interwar years was a rigid formula encouraging standardised suites of solutions. Harland Bartholomew's widely replicated formula of six heads of consideration for city planning had major streets, zoning, transit, and transportation complemented by the left-over culture-splitting categories of recreation and civic art.[12]

Some planners like the eccentric Patrick Geddes advocated more place-sensitive approaches. Occasionally the set formula could be adapted to acknowledge the character of particular places. H.V. Lanchester's (1916) civic survey for Madras in India expanded to include 'education and culture agencies', a broader appreciation

------

12 E. Lovelace, *Harland Bartholomew: His Contributions to American Urban Planning* (Champaign: University of Illinois, c1993).

of 'artistic problems in connection with civic design', local history and archaeology. More typically, such category-crossing or more innovative designations were denied.

The bifurcation of urban culture carried into the high modernism of Le Corbusier in the 1920s. His Radiant City represented a higher plane of interactive urbanity than captured in the garden city model, but remains in the social segregationist mode. At the base of the (business) skyscrapers would be restaurants, cafes, shops, theatres, and halls with great public buildings nearby. The workers living at the city's margins would have access to entertainments of 'a different sort, appropriate for those who worked hard for eight hours a day'.[13]

The civic survey formula at best allowing for a zoning of culture endured into the era of postwar reconstruction. Patrick Abercrombie's influential model summarised in *Town and Country Planning* promoted the same basic template over nearly three decades.[14] The survey step of survey-analysis-plan would encompass base-level investigations of physical features; history, archaeology and architecture; communications; industrial survey; population; health conditions; housing; open spaces; land utilisation; landscape survey; administration and finance, and public services. This was a conservative, status quo approach, with culture/creativity not yet written into the mix. However, such dimensions of urbanism might be acknowledged in specialised centers. Two of Abercrombie's own plans evidence this and endorsed planning on the 'precinctual idea'. His *County of London Plan 1943* recognised the spontaneous or 'natural zoning' of cultural districts such as Theatre and Cinema Land of the West End, the Soho Restaurant Area, and the 'Museum and Institution Centre' at South Kensington (the major aim is to protect them from through traffic). Abercrombie's *Plan for Bath* (1945) also envisaged the Georgian town conserved as 'a centre for the Arts', making the extraordinary suggestion of recycling the famous Royal Crescent into a civic centre.

Into the 1950s, cultural considerations in planning mainly meant the provision of appropriate facilities in an orderly and hierarchical manner. Central civic and highbrow facilities came first, then commercial entertainments tidily placed in ideal town center plans, followed by suburban community and neighbourhood centres providing for everyday needs (Figure 2.2). This form of urban cultural policy expressed 'very patrician forms of municipal provision'.[15] A new idea emerged in multi-purpose buildings offering flexible use opportunities for smaller communities. Culture was rarely seen as having a direct economic development rationale; it was more about public welfare. Governments responded in different ways. Some European and socialist states actively supported cultural programs for nation-

---

13　Hall, *Cities of Tomorrow: An Intellectual History of Urban Planning and Design in the Twentieth Century*, p. 25.

14　P. Abercrombie, J. Owens and H.A. Mealand, *A Plan for Bath* (Bath and District Joint Planning Committee, 1945).

15　G. Mulgan, and K. Worpole, *Saturday night or sunday morning* (London Comedia Publishing Group, 1986), p. 27.

Figure 2.2   Community centres were seen as engendering opportunities for cultural activity at the local level: Proposed centre in Melbourne (1944)

Source: F.O. Barnett, W.O. Burt and F. Heath, *We Must Go On: A study of planned reconstruction and housing* (Melbourne: The Book Depot., 1944).

building and national identity.[16] Others remained reluctant to spend big money on the arts although cold war thinking began to link the celebration of national high culture with international politics and foreign policy.

Urban Renewal and Flagship Facilities

Into the early 1960s, while cultural institutions were occasionally included in urban renewal schemes, 'most city planners and business–people still saw investments in culture as incidental to the main city development goals of industrial retention and office and housing development'.[17] One famous New York scheme well ahead of its time prefigured other possibilities. With its origins in Mayor Fiorello la Guardia's employment-generating promise of 'a popular municipal art center and conservatory', the high rise Rockefeller Center complex in midtown Manhattan from the 1930s was an extraordinary mixed-use development comprising a mix of commercial, cultural and public uses. Its key tenants were American cultural industry icons: NBC, RCA, RKO Pictures, and *Time Magazine*.[18]

---

16 E. Hitters, 'Culture and Capital in the 1990s: Private Support for the Arts and Urban Change in the Netherlands', *Built Environment*, 8 (1992): pp. 111–122.

17 E. Strom, 'Converting Pork into Porcelain: Cultural Institutions and Downtown Development', *Urban Affairs Review*, 38 (2002): pp. 3–21, p. 6.

18 V. Newhouse, *Wallace K Harrison, Architect* (New York: Rizzoli, 1989).

The Metropolitan Opera withdrew from the Rockefeller Center after the stock market crash but re-emerged as an anchor facility 30 years later in another influential New York icon, the Lincoln Centre for the Performing Arts. The *New York Times* welcomed this scheme in 1956 as 'the boldest and most exciting artistic project ever attempted in the United States'.[19] Three main buildings were completed in a major renewal exercise between 1962 and 1966: Avery Fisher Hall, State Theatre, and Metropolitan Opera House. Again Rockefeller money was brought into an alliance with other key city players including the New York Philharmonic, Juillard School for Dance, Fordham University, the urban 'power broker' Robert Moses of the New York Port Authority, and veteran architect Wallace Harrison, also the main designer of the Rockefeller Center.

Beyond Manhattan, new arts/corporate sponsorship/developer coalitions began to take shape in the 1960s laying the foundation for extensive construction of new, large facilities in the US and internationally. The trend was toward one-off heroic statements initiated outside any general plan. What was remarkable about many such projects was how little removed they were from the civic-cultural touchstones of the city beautiful era, often reinforcing these earlier developments. In New York, for example, the Lincoln Center recalls the Court of Honour in the Chicago World's Fair, is indebted to the symmetrical planning of Rockefeller Center, and in turn reveals the contemporary legacy of the beaux arts.[20]

Jane Jacobs made the most devastating and influential critique of this style of cultural spatial planning in the early 1960s.[21] She blamed it on a continuation of city beautiful allied to garden city thinking. Monofunctional urban districts were 'a pitiful kind of planning', 'woefully unbalanced' and 'tragic' in their effects, undermining urban vitality by segregating and 'isolating' uses. They represented an ill-conceived policy of 'decontamination', disconnecting cultural institutions from the fabric of the existing city. Jacobs was particularly caustic in her criticism of the oppressive planning logic of 'sorting out' embodied in the Lincoln Center development (Figure 2.3). Other concerns were also expressed – of a development insensitively imploded into yet disconnected from the existing city, the reliance on federal housing funds to subsidise elite artistic activity, and the displacement of thousands of families. The local community was stirred into unsuccessful protest, albeit a bellwether of more sustained community actions to follow.

---

19 J. Schwartz, *The New York Approach: Robert Moses, Urban Liberals, and Redevelopment of the Inner City* (Columbus: Ohio State University Press, 1993), pp. 280–281.

20 Newhouse, *Wallace K Harrison, Architect.*

21 J. Jacobs, *The Death and Life of Great American Cities* (New York: Vintage, 1961).

Figure 2.3   Flagship cultural institution of the 1960s: Avery Fisher Hall in the Lincoln Center for the Performing Arts in New York City (1976)
Source: R. Freestone.

The Cultures of Communities

The 1960s and 1970s were decades of massive social change. Emergent was a more democratic concept of culture denoting not just the practices and values of high art but whole ways of life and behaviours in an almost anthropological sense. Art itself began to lose its *bourgeois* aura. In western cities, these fractures came at a time of student dissent, anti-war protests, environmentalism, and civil rights activism. The critical questioning of the universal commitment to what Ward[22] calls 'technocratic and comprehensive urban modernisation' had serious implications for planning at this time. The monolithic rational-comprehensive planning paradigm was challenged in parallel and began to fragment into multiple specialisms.[23] There was no longer one 'planning', just as there were multiple 'cultures', and the possibilities of their cross-fertilisation accordingly expanded.

At the local level, an early manifestation was the community arts movement. In Britain, for example, new arts centers were opened in existing buildings or in

---

22 S. Ward, *Planning the Twentieth-Century City: The Advanced Capitalist World* (New York: John Wiley, 2002), p. 269.

23 O. Yiftachel, 'Towards a New Typology of Urban Planning Theories', *Planning and Environment B: Planning and Design*, 16 (1989): pp. 23–39.

new multipurpose structures offering 'a more contemporary and less institutional setting for arts participation'.[24] Novel avenues were opened up for self-expression, the genesis of new creative gestures, cultural jobs, community participation, and social critique. The very nature of cities began to change as new meanings and appreciations of culture were liberated. A crucial element worldwide was the re-evaluation of the inherited built environment and development of strong historic preservation and heritage conservation movements. Saving neighbourhoods from urban renewal schemes and adaptively reusing abandoned structures and precincts went beyond immediate aesthetic and amenity benefits to stabilise (then inflate) property values, promote the arts economy, and attract new commercial and public investment.

The pro-active responses of planning systems varied. One direction was for mainstream methodology to develop new standards of provision, arts plans and cultural audits to inventorise needed facilities of different types and scales. Second were pioneering economic and employment impact studies of the arts and cultural industries (e.g. Basle in 1976; Baltimore 1977; Vancouver 1976). Third, came greater inventiveness in the 'softer' and low cost areas of social and cultural planning, even when high culture ruled in the dominant physical realm. Fourth was a greater engagement of community arts and social action movements with urban policy initiatives. In Europe, the more innovative cities had first felt the impact of declining manufacturing and traditional port functions from the early 1970s came under the control of progressive 'new left' administrations.[25]

Wider possibilities for the 'diffusion of art and culture in state-financed programs of community and economic redevelopment' were recognised.[26] Competing growth engines arrived in waves of popularity: conventions, exhibition centres, sports stadia, casinos. In the 1970s 'shopping and dining' were seen as crucial instruments.[27] Two key mid-1970s projects updating the flagship idea within a more fluid conception of commercial culture and planning helping navigate global development trends came in Boston and Paris. James Rouse's Quincy Market (1976) in Boston recycled derelict buildings for specialist and leisure retailing in a pedestrian precinct to become the prototype for the festival marketplace. In the French capital, completion of the Pompidou Centre set off a worldwide boom in new and renovated art museums.[28]

---

24 Evans, *Cultural Planning: An urban renaissance*, p. 94.

25 Ibid.

26 Zukin, *The Cultures of Cities*, p. 121.

27 J. Hannigan, *Fantasy City: Pleasure and profit in the postmodern metropolis* (London: Routledge, 1998).

28 I. van Aalat and I. Boogaarts, 'From Museum to Mass Entertainment: The Evolution of the Role of Museums in Cities', *European Urban and Regional Studies*, 9 (2002): pp. 195–200.

Culture in Urban Development

The 1980s and early 1990s marked the most dramatic and decisive cultural turn in planning. A crucial catalyst was the negative urban impacts of economic restructuring. The collapse of hegemonic Fordism set in train processes of deindustrialisation and the search for new sources of investment, employment-generation, and comparative advantage place-making. A new economic environment of globalisation was matched by the lurch to pro-market and small government neo-liberalism. In this new climate, the economic importance of the arts was given yet more recognition. The 'culture-generating capabilities of cities [were] being harnessed to productive purposes'.[29] Accompanying and interpenetrating this process was a rising 'anything goes' postmodernist take on culture.

Bassett[30] identifies four propelling factors in the UK. Thatcherism saw state patronage for the arts cut back and funding linked to new privatisation and devolutionary models. New left urban politics were grounded in cultural populism and pluralism. Changes in the construction and consumption of culture further transcended the traditional cleavage between highbrow and lowbrow, introducing a new breed of professional 'cultural intermediaries'. Finally, rapid economic restructuring would be ultimately played out in greater urban entrepreneurialism and city 're-imaging'.

An 'urban renaissance' of sorts became evident, one incorporating culture variously as a 'consumption, production and image strategy'.[31] All scales of government were impacted, from the local to the supranational (perhaps best captured at this time by the European City of Culture program launched in 1985 with Athens). From flagship projects and hallmark events to grassroots participation and local festivals came a wider appreciation of the significance of cultural infrastructure and cultural diversity in local economic development. The suite of policies included the often interrelated promotion of:

- culture-led urban regeneration
- cultural tourism
- place marketing, city re-imaging and civic boosterism
- cultural districts
- cultural industries
- community-based cultural planning.

In turn these were advanced by a mix of specific strategies and agencies:

- public-private partnerships

---

29 Scott, *The Cultural Economy of Cities*, p. 14.

30 K. Bassett, 'Urban cultural strategies and urban regeneration: a case study and critique, *Environment and Planning A*, 25 (1993): pp. 1773–1788.

31 Evans, *Cultural Planning: An urban renaissance*, pp. 1–2.

Figure 2.4   Urban cultural icon of the early twenty-first century: Walt Disney Concert Hall, Los Angeles (2004)
Source: R. Freestone.

- special development incentives
- mixed-use developments
- tapping new funding sources, e.g. lotteries
- public relations
- arts bodies, not-for-profit organisations, foundations and trusts
- consultative and participatory exercises at the local level
- private planning.

The responses have been categorised as mainly either consumptionist or productionist, complemented by social integrationist and artist-led initiatives.[32] The emphasis on cultural consumption has underpinned the most notable exercises in culture-led urban regeneration. In the United States, for example, into the early 1990s and beyond cities were in pursuit of 'the big 4': an orchestra and concert hall, a performing arts company with matching performance space, a national museum, and a major visual arts center.[33] The performing arts were increasingly seen as a 'new fuel for urban

32 R. Griffiths, 'Urban planning and cultural development: the role of artist organisations', in R. Freestone (ed.), *The Twentieth Century Urban Planning Experience, Proceedings of the 8th International Planning History Society Conference* (University of New South Wales, 1998), pp. 277–282.

33 Comedia, *Glasgow: The creative city and its cultural economy*, p. 27.

growth machines'[34] and since 1985 well over 70 new major performing arts centres and museums have been constructed (Strom, 2002). The new Walt Disney Concert Hall in the Los Angeles Music Centre Complex in downtown Los Angeles is yet another Frank Gehry design to attract international attention.

Two key factors behind this 'urban cultural building boom' were cities seeking to increase their symbolic capital through quality-of-life amenities to attract business as well as cultural institutions driven by their own financial imperatives toward greater income-generating activities.[35] A raft of specific drivers and catalysts has facilitated this process, and not only in the US, including historic conservation, the revival of downtown commercial cores, displacement of locally undesirable uses (LULUs), and promotion of social inclusiveness.[36]

At the same time arose a more critical appreciation of the cultural industry production in economic development. During the 1980s and 1990s:

> Cities across Europe became 'more and more preoccupied by the notion that cultural industries .... may provide the basis for economic regeneration, filling the gap left by vanished factories and warehouses, and creating a new urban image that would make them more attractive to mobile capital and mobile professional workers.[37]

Cultural industries exemplify trends toward flexible specialisation, customisation, market segmentation, and just-in-time production in the new economy. While the common description of juxtaposed cultural facilities as 'cultural quarters' has been dismissed as a mere 'branding device',[38] the clustering of complementary cultural producers from pre-industrial crafts to new media industries (e.g. broadcasting, music, design, multimedia) has been promoted by some cities and state planning authorities as an extension of natural economic agglomerating forces. The urban policy response (extending beyond traditional land use planning) has encompassed infrastructure for transport, skills training, and workplace amenities.[39]

---

34 J.A. Whitt, 'Mozart in the Metropolis: The Arts Coalition and the Urban Growth Machine', *Urban Affairs Quarterly*, 23 (1987): pp. 15–36, p. 16.

35 Strom, 'Converting Pork into Porcelain: Cultural Institutions and Downtown Development'.

36 J.A. Whitt and J.C. Lammers, 'The Art of Growth: Ties Between Development Organizations and the Performing Arts', *Urban Affairs Quarterly*, 26 (1991): pp. 376–393.

37 Hall, *Cities in Civilisation: Culture, Innovation and Urban Order*, p. 8.

38 C. Landry, The Creative City: A Toolkit for Urban Innovators (London: Earthscan, 2000).

39 D. Bell and M. Jaynes (eds.), *City of Quarters: Urban Villages in the Contemporary City* (Ashgate: Aldershot, 2004); Evans, *Cultural Planning: An urban renaissance*; and S. Yusuf and K. Nabeshima, 'Creative industries in East Asia', *Cities*, 22 (2005): pp. 109–122.

Figure 2.5 Cultural thinking infusing urban planning: The Creative City strategy in Brisbane (2003)

Source: Brisbane City Council (reproduced with permission).

The Creative City

At a macro-level we can now see an historic evolution from a manufacturing to an informational to a cultural economy fuelled by forces of global capital, international tourism, and the search for comparative economic advantage. Sharon Zukin announced as early as 1995 the arrival of 'an artistic model of production'. Not only is the association between development policy and culture 'commonplace', cultural institutions are no longer just for connoisseurs but are frequently 'an explicit part of a city's economic revitalisation program'.[40] Art, culture and entertainment are increasingly employed as 'instruments of spatial planning'.[41]

Urban economies increasingly dependent on the production and consumption of culture signal the arrival of the 'creative cities' paradigm.[42] Charles Landry[43] casts culture as the raw material of the city and creativity as the means of exploiting this resource. Although 'creative city' rhetoric has quickly become almost a 'civic boosterist cliché',[44] it still captures a new zeitgeist at the interface of culture, planning and urban development with significant new overlaps and relationships between stakeholders in these sectors (Figure 2.5). The notion is strengthened by American guru Richard Florida's[45] thesis that place–based creativity is now the fundamental source of economic growth.

Actual forms of cultural-based planning often remain a continuation and adaptation of 1980s models. There is now an enormous variety of activities in which planning is deeply and increasingly implicated. The partial list developed by Evans is indicative: arts and media centres, theatres, museums and galleries, centres for *edutainment*, public gatherings, raves and festivals, 'Pavarotti in the Park' type-concerts, public art works, urban design and public realm schemes, promotion of cultural industries zones and workplaces, and sundry grand projects.[46]

Cultural led regeneration is now pervasive. As one indicator, over 40 city projects of this kind were presented at the 2001 Biennial of Towns and Town Planners in Europe. Their great variety is unified by recognition of the transformative qualities of urban revitalisation, the integrative benefits of coordinated urban design, and the growing importance of spatial planning throughout Europe.[47] Triple-bottom line assessments of the social, environmental and economic impacts of cultural element

---

40 Strom, 'Converting Pork into Porcelain: Cultural Institutions and Downtown Development', p. 5.

41 I. van Aalat and I. Boogaarts, 'From Museum to Mass Entertainment: The Evolution of the Role of Museums in Cities', *European Urban and Regional Studies*, 9 (2002): pp. 195–200, p. 196.

42 P. Hall, 'Creative Cities and Economic Development', *Urban Studies*, 37 (2000): 539–49.

43 Landry, *The Creative City: A Toolkit for Urban Innovators.*

44 D. Stevenson, *Cities and Urban Cultures* (Maidenhead: Open University Press, 2003), p. 104.

45 R. Florida, R., *The Rise of the Creative Class: And How It's Transforming Work, Leisure, Community and Everyday Life* (New York: Basic Books, 2002); and R. Florida, *The Flight of the Creative Class: The New Global Competition for Talent* (New York: Harper Business, 2005).

46 Evans, *Cultural Planning: An urban renaissance*, p. 2–3.

47 D. Bayliss, 'Creative planning in Ireland: The role of culture-led development in Irish planning', *European Planning Studies*, 12 (2004): pp. 497–515; J. Vogelij, 'Introduction to the Exhibition', *Cultures of Cities: A new data bank*. Planum – 4th Biennial (Rotterdam: 2003), *www.planum.net/4bie/main/m–index.htm*; and M. Wansborough and A. Mageean, 'The Role of Urban Design in Cultural Regeneration', *Journal of Urban Design*, 5 (2000): pp. 181–197.

in urban regeneration remain somewhat elusive.[48] Certainly, it has become clear that such projects cannot be 'one-offs'; they need to be part of cyclical and integrated strategies. Hence, Bilbao's famous Guggenheim Museum (1997) is but one element of a broader program of economic growth, infrastructure development, and upgrading of cultural amenities. Glasgow, an early 1980s icon, is now in its third phase of cultural regeneration programming. Just like retail centers, cultural facilities and institutions must experience sequential processes of refurbishment, renovation and re-invention to keep in touch with new market demands.

Cultural institutions (from theatres to film studios) are figuring more and more in urban development projects, even in speculative ventures in return for extra commercial floorspace and other concessions. When New York's World Trade Centre was conceived in the 1960s, cultural uses were 'not in the picture'. It is a very different picture post-September 11 which brought home to the city 'the phenomenon of urban renewal through culture'.[49] Cultural institutions themselves have become 'active promoters of revitalisation and place marketing' in recognizing the interdependence of economic health, institutional citizenship, fund-raising, community amenity, and tourism.[50] Their ability to do that has obviously been enhanced by their genre-crossing, crowd-pulling evolution away from highbrow elitism.

The expression 'cultural planning' has been attributed to American planner Harvey Perloff in his *The Arts in the Economic Life of the City* (1979) as a way for communities promote both artistic excellence and community building.[51] Stevenson writes that 'city-based cultural planning has emerged to be the most significant local cultural policy innovation of recent years'.[52] It has developed in various guises and at different scales. At the local level, for instance, it is seen as a means of enhancing participatory democracy and citizenship, although sensitively and equitably dealing with seedbed creative spaces (run-down districts, warehouses, squats) largely remains elusive. Cultural audits and strategies are proving popular with municipal authorities who are also increasingly turning to 'cultural directors' with a more entrepreneurial market-friendly approach to local cultural development.

Summation

What of other emergent trends and needs? Projecting some key developments already evident, we might expect:

---

48 G. Evans, 'Measure for Measure: Evaluating the Evidence of Culture's Contribution to Regeneration', *Urban Studies*, 42 (2005): pp. 959–983.

49 J. Salamon, 'Making a fine art of urban renewal', *Sydney Morning Herald*, 17 September 2003 (from *New York Times*).

50 Strom, 'Converting Pork into Porcelain: Cultural Institutions and Downtown Development', pp. 9–10.

51 G. Dreeszen, *Community Cultural Planning: A Guidebook for Community Leaders*, (Americans for the Arts, 1998).

52 Stevenson, *Cities and Urban Cultures*, p. 104.

- private planning
- still greater hybridity and diversity in constructions of culture
- intensification of the market-oriented environment for cultural industries and institutions
- greater attention to sense-of-place and place management strategies
- rising importance of urban design and public art
- even more entrepreneurial competitiveness between planning jurisdictions for scarce cultural resources
- suburbanisation of large-scale cultural institutions
- more systematic integration into formal planning strategies at all scales, in particular, within regional and suburban initiatives
- greater interconnection of culture into other strategies, e.g. transport, infrastructure, mix-use developments
- shifting to productionist initiatives and greater role of private sector
- orientation toward strategic new vehicles for culture advancement, e.g. campuses
- more systematic codification and diffusion of cultural planning primers and methodologies
- greater embrace of digital technologies in mapping local cultural resources.

We cannot explore the ramifications of these developments (not all benign) given our historic focus; that is underway elsewhere.[53] Suffice to say that the overarching implication is that the cultural turn in planning is so pronounced that some commentators have already proclaimed that urban policy is inseparable from cultural policy:

> Any form of urban planning is today, by definition, a form of cultural planning in its broadest sense, as it cannot but take into account people's religious and linguistic identities, their cultural institutions and lifestyles, their modes of behavior and aspirations, and the contributions they make to the urban tapestry.[54]

Charles Landry's[55] toolkit for the creative city in effect has been cast as a new method of strategic urban planning – at once holistic, open-minded, inventive, and alive to non-tangible planning resources. In these terms, the scope of cultural

53 F. Bianchini and M. Parkinson (eds), *Cultural Policy and Urban Regeneration: The West European Experience* (Manchester: Manchester University Press, 1993); G. Evans and J. Foord, 'Shaping the Cultural landscape: Local Regeneration Effects', in Malcolm Miles and Tim Hall (eds), *Urban Futures: Critical commentaries on shaping the city* (London: Routledge, 2003), pp. 167–181; and K.R. Kunzmann, 'Culture, creativity and spatial planning', *Town Planning Review*, 75 (2004): pp. 383–404.

54 K. Worpole and P. Greenhalgh, *The Richness of Cities: Urban Policy in a New Landscape*, (London: Comedia/Demos, 1999), p. 4.

55 Landry, *The Creative City: A Toolkit for Urban Innovators.*

planning becomes so vast that it can be portrayed as 'a new way of approaching urban planning'.[56]

Conclusions

Culture is a most instructive vector for planning histories. The twentieth century was characterised by a reconceptualisation of culture away from something that was 'bounded' and 'monolithic' to something 'more fluid and permeable'. Where once city and culture were distinct, they are now 'inextricably intertwined and mutually influential'.[57] Responding to fundamental shifts in the economic, competitive, demographic, technological and environmental bases of urban society, both culture and planning have trended in the same direction.

For much of the twentieth century, the cultural basis of planning reached out selectively to high culture, privilege, and citadel-monumentalism. It was not quite 'art for art's sake', but culture was valued more for its educational, civilising and ennobling values than its economic credentials. Culture was zoned spatially and symbolically as a by-product of economic surplus rather than an instrument in its creation. Complementing this elite orientation was a facility-provision model addressing local community needs. The sea-change came in the late 1960s. Category-busting notions of culture emerging from local community action as well as an urban fiscal crisis spill over into a growing awareness of the economic impact of the arts. Progressive authorities begin to initiate audits, plans and strategies to promote local economic development via cultural industries. This in turn laid the foundations for integrated cultural infused urban regeneration strategies. From the early 1980s, diverse forms of cultural planning and development ensued as part of new urban renaissance 'in order to inform the formulation of policies and strategies, cutting across the public and private sectors, different institutional concerns and differing professional disciplines'.[58] Latterly the 'creative city' paradigm has taken this to a new level wherein culture is portrayed as the foundation for planning. A century apart, the notion of 'creative cities' bookends 'garden cities' as a defining planning paradigm.

We are not suggesting a singular narrative everywhere. While this chapter has attempted to set the scene via a chronological collage of ideal-typical trends, it does not suggest this is either the story of all places or any one place. As well as the strong convergences highlighted in this overview, there have clearly been distinctive variations and practices because local forces dictate 'the nature and scope of cultural city formation'.[59] Studies of cities in their international setting inevitably highlight a raft of factors promoting divergence to norms – place-specific histories, planning

---

56 Stevenson, *Cities and Urban Cultures*, p. 105.

57 N. Ellin, *Postmodern Urbanism* (Oxford: Blackwell, 1996), pp. 247 and 250.

58 Comedia, *Glasgow: The creative city and its cultural economy*, p. 24.

59 Evans, *Cultural Planning: An urban renaissance*, p. 215.

cultures, and governance systems. Nations, regions, cities and localities are not 'mere leaves in the wind, passively responding to international change'.[60]

The international diffusion of ideas is a central concern of planning history and cultural planning presents a rich field for further exploration. Although the early pace of change was glacial, there was clearly a sharing of ideas and convergence around a similar conservative model. Global cities became and remain influential lighthouses of innovation.[61] From the 1970s US downtown cultural revitalisation led by James Rouse's festival marketplace formula proved very influential, as did European urban regeneration and city reimaging in the 1980s. The creative interplay between these dominant late twentieth century paradigms has been played out in many places. In the new planning history, global culture city exemplars (Bilbao, Baltimore, Boston, Barcelona) are now name-checked as planning successes eclipsing the icons of earlier eras (Chicago, Canberra, Crawley, Chandigarh). Latterly, Stevenson observes 'an intriguing global circuit of cultural planning consultants selling a "just add local culture and stir" version of cultural planning', perhaps a thinly veiled reference to Comedia.[62] Yet, for our purposes, the evolution of this company registers the rapid rate of change of the cultural-planning nexus in recent years. Since the 1970s it has evolved from a campaigning collective assisting the organisation and marketing campaigns of radical groups through feasibility studies for arts and cultural projects, investigations of the potential of cultural industries, and the linking of broader ideas on art and culture to the city, toward a new inclusive notion of creative power in urban economic development.[63]

The spatial imprint of planning has been another enduring concern of contemporary planning history. From the early twentieth century the historical geography of urban cultural development appears to have been remarkably consistent. Generic 'east–west' and 'city–suburban' contrasts in cultural resource provision and access evident from the nineteenth century have largely been sustained.[64] But latterly, abandoned brownfields, waterfront and other inner city areas have become the dominant focus of urban regeneration everywhere. The unglamorous and left-over spaces of the Fordist city have become central to the post-industrial city.[65]

While it now comes as a shock to see culture excluded from major strategic and spatial planning exercises, to claim like some commentators that culture has risen to become define the essence of planning both overstates the centrality of planning

---

60 H.V. Savitch and P. Kantor, *Cities in the International Marketplace: The political economy of urban development in North American and Western Europe* (Princeton: Princeton University Press, 2002), p. xviii.

61 J.R. Short, *Global Metropolitan: Globalizing Cities in a Capitalist World* (London: Routledge, 2004).

62 Stevenson, *Cities and Urban Cultures*, p. 106.

63 N. Camilleri, *The Creative City: Reality or Rhetoric? A critique of the recent work of Charles Landry and COMEDIA*, BTP thesis, (University of New South Wales, 2001).

64 Evans, *Cultural Planning: An urban renaissance.*

65 D. Wynne, *The culture industry: The arts in urban regeneration* (Avebury: Ashgate, 1992).

in cultural life and marginalises the extraordinary range of other activities which must still be assimilated by planning activity at different scales. We can safely conclude that culture has clearly become yet another social dimension for planning to juggle into its remit of competing concerns. Simultaneously, planning has to deal with culture not only as an integral part of urban economies but as a framework for community development and societal engagement. It has to negotiate and better bridge the divide between the conservative property focus of physical planning and the more insurgent demands of social planning in the search for more inclusionary methodologies. It has to respond to postmodern utopianism's call for an anti-bureaucratic celebration of 'the carnival of the multicultural city'[66] and at the same time accommodate the humdrum of regulation and development control. It has to successfully juxtapose the need for more integrated cultural planning strategies with an ingrained view within the creative sector that culture is inherently damaged when centrally planned and administered.[67] To what extent can planning play a decisive role when mess and disorder are seen as crucial creative incubators? Peter Hall[68] provides a succinct but salutary 'lesson for aspiring planners' in this ruling age of the creative city: 'Building innovative milieux is not something that can be done either easily or to order'.

Acknowledgements

Our thanks to Graeme Evans, Lisanne Gibson, and Charles Landry for their comments on an earlier draft of this chapter.

---

66 L. Sandercock, *Cosmopolis II: Mongrel Cities of the 21st Century* (New York: Continuum, 2003), p. 207.

67 T. Bennett, *Culture: A Reformer's Science* (Sydney: Allen and Unwin, 1998).

68 Hall, *Cities in Civilisation: Culture, Innovation and Urban Order*, p. 498.

Chapter 3

Speak, Culture! – Culture in Planning's Past, Present and Future

Greg Young

Greg Young is a Sydney-based planner and historian, who has prepared a number of landmark Australian strategies for culture, heritage and tourism, including the pioneer Australian model for cultural mapping. The inaugural NSW Premier's Max Kelly Scholar to Venice, Italy, Dr Young consults and writes internationally. He is currently preparing a book on his concept of the practical culturisation of planning in its urban and regional and non-spatial strategic forms.

In the volume *Speak, Memory* (1969),[1] the autobiographical recollections of the novelist Vladimir Nabokov are presented, saturated with the history and qualities of the famous writer's life and times. In his choice of a brusque and unforgettable title Nabokov announces both an arresting purpose and a key metaphor for his undertaking. In this article 'Speak, Culture', I wish to suggest in a similar decisive fashion, a rationale and an approach for another project – the effective integration of culture into planning. This is long overdue, for lamentably, while culture is embedded in geographies, societies and histories, its voice is weak in planning. In fact culture rarely seems to speak meaningfully in planning at all. Although the reasons for this are numerous, perhaps the most fundamental is the fact that the power of culture as an organising concept has only lately begun to be perceived, or utilised. And then, by its very nature, culture appears to resist a rigorous and ethical integration into planning. Examples of the systematic and foundational inclusion of culture in urban and regional planning are in short supply all around the globe. What familiarity with culture there is, tends to lie more with its limited and often opportunistic inclusion in strategic planning and specific planning sectors, such as tourism, heritage, and marketing. This inclusion reflects quite directly a broader, emerging picture – that of culturalisation. This is the complex pattern of an era 'in which economic and organisational life has become increasingly "culturalised"'.[2] The reason for this is the fact that under modern capitalism, cultural production is 'increasingly commodified

---

1   V. Nabokov, *Speak, Memory* (London: Penguin Books, 1969).
2   P. du Gay and M. Pryke (eds), *Cultural Economy: Cultural Analysis and Commercial Life* (London: Sage, 2000), p. 6.

while commodities themselves (are) increasingly invested with symbolic value'.[3] This is particularly manifest in the growth of the cultural economy, encompassing for example advertising, marketing, the arts, and the media. In the field of the arts as Evans notes culturalisation can include cultural planning and planning for the arts, particularly in terms of 'the growing attention paid to the cultural economy and the commodification of the arts as urban cultural assets'.[4] Broader planning follows this trend especially in sectors such as development, tourism and heritage, where cultural forms and cultural content are increasingly incorporated in planning and commodified in the process.

In contrast to the many manifestations of the trend to culturalisation, I propose a radically different alternative for planning that may enable culture to speak more on its own terms. This alternative I tag with the new name of 'culturisation'. Culturisation I define as the systematic research and ethical and reflexive integration of historical and contemporary cultural knowledge, theory, and interpretation in spatial and strategic planning. It also potentially applies to other social technologies. I also explore its planning antecedents and its possible future. Before undertaking this journey however I suggest some of the qualities and characteristics of contemporary culture and the fashion in which it now operates and has come to be experienced. This is the background to any successful culturisation of planning, based on an understanding of the new position of culture and a new positionality for planning based upon it.

Culture Today

The voice of culture is acquiring unprecedented authority and interest in planning today. This is reflected in the increasing inclusion of the multiple dimensions of culture in planning practices across the entire spectrum of planning experience. These dimensions encompass the tangible and intangible forms of culture and culture's historical and contemporary manifestations. Nevertheless the inclusion of culture in planning today is a consistently uneven feature of international planning experience. At the same time, culture's global role as a new fulcrum for societies, economies and environments, is the basis of a remarkable opportunity for planning. Further, it is an approach that is widely advocated. United Nations agencies, advisers to these bodies and numerous theorists and commentators march to the same tune of cultural integration.[5] In the meantime, the leaking holes in the residues of modernist

3   A. Scott, *The Cultural Economy of Cities – Essays on the Geography of Image-Producing Industries* (London: Sage Publications, 2000), p. 3.

4   G. Evans, *Cultural Planning. An Urban Renaissance?* (London: Routledge, 2001), p. 16.

5   E. Soja, *Postmodern Geographies. The Reassertion of Space in Critical Social Theory* (London: Verso, 1993); World Commission on Culture and Development, *Our Creative Diversity* (Paris: UNESCO, 1995); L. Sandercock, *Towards Cosmopolis – Planning*

planning practice are being plugged with fresh cultural content and, more importantly, planning practices and planning systems themselves are being slowly transformed. For example, in relation to the growth and prominence of cultural diversity, planning is recognising and responding to this trend with a new breadth and relevance in social planning, heritage planning, and tourism planning. It is also impacting on the design and development sectors, and in sustainability approaches, as reflected in the cultural perspective of Agenda 21 of the 'Rio Summit'.[6] Simultaneously, cultural diversity itself is splintering into ever-new cultural fractions, especially in the global, or 'mongrel cities' discussed by Sandercock.[7] These cultural fractions are expressed across ethnicities, arts practices, and consumer goods representing an interrelated mosaic of hybridity.

What this collective global experience suggests is that the well-documented 'cultural turn'[8] of the late twentieth century has evolved further into the new totality of culturalisation. Culturalisation is now the essential nature of contemporary societies, economies and signification, and impacts on planning as well as the workings of culture in their entirety. It is not only transforming societies and their planning but, for good measure, the face of the planet. While this represents a new opportunity for planning, it is also a threat as I will come to describe. The opportunity and the threat apply both for urban and regional planning and for strategic planning in its non-spatial forms. However, why and how this has come to pass is first in need of further explanation, and I begin with the experience of the twentieth century.

Culture and Planning in the Twentieth Century

In the United Kingdom and Australasia, prior to the Second World War and particularly the watershed period of the 1970s, a dominant concept of culture based on the concept of high culture reigned. With the onset of the 1960s and 1970s a rival concept of culture based on culture as 'a whole way of life',[9] defined in the writings of Raymond Williams, was introduced and gained purchase. This emergent concept influenced society and planning increasingly and is now dominant. However, throughout this period the two concepts of culture reflected in their rivalry, distinctive cultural and social bases and interests, and were differentially integrated into planning with little awareness of contradiction. Planning was slow to represent the onset of new cultural ideas in overall terms and tended to exhibit a form of conceptual schizophrenia. A dominant or declining idea of culture was uppermost in planning's 'mainstream'

---

for Multicultural Cities (Chichester England: John Wiley & Sons, 1998); L. Sandercock, *Cosmopolis II. Mongrel Cities in the 21st Century* (London: Continuum, 2003); and Scott, *The Cultural Economy of Cities – Essays on the Geography of Image-Producing Industries.*

6   United Nations, *United Nations Conference on Environment and Development, Agenda 21* (New York: UN Dept. of Public Information, 1993).

7   Sandercock, *Cosmopolis II. Mongrel Cities in the 21st Century.*

8   D. Chaney, *The Cultural Turn* (London: Routledge, 1994).

9   R. Williams, *Culture and Society, 1780–1950* (England: Penguin, 1966).

land-based activities – the statutory and spatial sectors with a close nexus to power and capital. More up to date ideas about culture have been influential on planning's perceived margins of social and 'cultural' planning and suchlike.[10]

Raymond Williams' introduction of a concept of culture as a way-of-life and the anthropological concept emphasising the need to interpret cultural variation, in retrospect represented seminal changes in ideas. These concepts of culture enabled the culture and values that inhered in working class lives, in indigenous cultures and in residual centres to be taken into consideration and considered and valued as important. This concept of a way-of-life was pluralised and extended into minorities represented by sub-cultures and ethnic, migrant and gay and lesbian communities. Further, the concept of ways of life and the anthropological 'interpretive' view of culture are both normative tools for the development of culture and planning.

At the global level, the United Nations resourced intellectual, conceptual and ethical work. In the field of culture, despite the cynicism of many commentators, work in conceptualising and creating standards for many aspects of culture has moved to the point where it now can be used to assist in defining the opportunities for planning, as I hope to have illustrated. The gap that lies in the area of creating supporting and enabling methodologies to realise the conceptual vision and the opportunities framework, however remains.

The concept of culture understood 'as ways-of-life' under the umbrella doctrine of culturalism, now dominates contemporary cultural discourse. The implication of these concepts for planning are being slowly realised as the de-cultured approach of planning modernism loses fails the test of relevance in the face of the ongoing expansion of the cultural economy[11] and the accelerating diversity of postmodern communities. The post-colonial emergence of indigenous and minority cultures and the powerful rethinking of historical culture are part of this. Added to this, are developing views from the late twentieth century that emphasise, firstly, the foundational nature of culture for all development and planning and, secondly, the contemporary fashion in which culture operates and expands through itself in a postmodern age.

Culture and governance

Woven into the preceding history of the understanding of culture in the period since the Second World War are the important layers of UNESCO philosophy and documents on culture, its forms, social priorities and uses. The United Nations founding *Universal Declaration of Human Rights*[12] occupies a key, catalytic role

---

10 G. Young, 'Concepts of Culture in Society and Planning in 20th to 21st Century Australia and Britain', *Planning History*: IPHS (forthcoming) (2005).

11 Scott, *The Cultural Economy of Cities – Essays on the Geography of Image-Producing Industries.*

12 United Nations, *Universal Declaration of Human Rights* (1948) <http://www.un.org/Overview/rights.html> viewed 1 March, 2005.

here. This Declaration identified cultural rights as defining rights of humanity and UN documents and those of its agencies since then have consistently promulgated the importance of culture and cultural rights. These concepts are crystallised in a framework of conventions and recommendations that continue to evolve.

Throughout the period, UNESCO documents link culture and human rights in an impressive fashion. For example, despite its unwieldy and patronising title, the *Recommendation on Participation by the People at Large in Cultural Life and their Contribution to it*, of 1976[13] reveals an integrated philosophy. The Recommendation stresses:

> the need to promote cultural rights as human rights on the grounds that, if cultural rights are recognised as human rights, a government is duty bound to establish the conditions required for the exercise of these rights and to frame cultural policies for that purpose.[14]

UNESCO's early normative struggles, based in rhetoric, laid the foundations for later developments. The concept of sustainability was integrated into planning at the Rio Summit and Agenda 21 nominated local cultural awareness as the foundation for the practical implementation of sustainability strategies for cultures and their environments. In the 1990s a number of the strands of earlier thinking coalesced in the 1995 findings of the World Commission on Culture and Development and its report, *Our Creative Diversity*.[15] The independent World Commission on Culture (WCC) was established in 1992 by the United Nations and its document represents the full flowering of culturalism at the level of international governance. It argues that cultural policy when considered as the basis of development must be understood in the broadest of terms. It should not only be sensitive to culture but inspired by it. This includes making better use of the realities of pluralism by finding the elements of cohesion that hold multi-ethnic societies together.[16]

Again I will illustrate a pattern thorough the use of a national example. Cultural heritage discourse is a global phenomenon and the development of heritage policy in Australia has been intertwined with the *Charter of Venice* (1964)[17] on conservation and the *World Heritage Convention*.[18] The pilgrimage of heritage discourse in Australia since the 1970s in Australia has been an important cultural journey. This

---

13 UNESCO, *Recommendation on Participation by the People at Large in Cultural life and their Contribution to it* (1976), viewed 1 March, 2005 (1976) <http://portal.unesco.org/en/ev.php-URL_ID=12026&URL_DO=DO_TOPIC&URL_SECTION=-471.html>.

14 G. Young, 'Critical Issues in Heritage and Conservation', unpublished MA Thesis (Sydney: University of Sydney, 1988), p. 30.

15 World Commission on Culture and Development, *Our Creative Diversity.*

16 Ibid.

17 ICOMOS, *International Charter for the Conservation and Restoration of Monuments and Sites* (Charter of Venice) (1964), viewed 1 March, 2005. <http://www.international.icomos.org/charters/venice_e.htm>.

18 UNESCO, *Convention Concerning the Protection of the World Cultural and Natural Heritage* (1972) <http://whc.unesco.org/world_he.htm> viewed 1 March 2005.

**Figure 3.1   Architectural culture respected as the city evolves. T & G Building,
Melbourne, Australia**

Photo: Australia ICOMOS.

story commences with the adoption in 1979 by the Australia International Council on
Monuments and Sites[19] of *The Australia ICOMOS Guidelines for the Conservation of
Places of Cultural Significance* (the 'Burra Charter'). This drew on the 1964 *Charter
of Venice*. The 'Burra Charter' was a major influence in establishing a more consistent
process for preparing conservation plans, based on a sequence of assessing cultural
significance, developing a conservation policy and strategy and implementing the
strategy. However, it was not until the 1990s that community values and cultural

---

19 Australia ICOMOS, *The Australia ICOMOS Charter for the Conservation of Places
of Cultural Significance* (Australia ICOMOS, 1999).

diversity were accorded a fuller place in heritage policy. An illustrated version of the 'Burra Charter' is currently published[20] and highlights the Charter's application to a range of subtle issues that selectively include the coexistence of cultural values, the relation between places and objects, conserving uses, retaining associations and meanings, adding new layers the fabric of a place, and respecting city streetscape. In the case of respect for city streetscapes, Figure 3.1 shows an important street in Melbourne that is depicted in *The Illustrated Burra Charter*. The 40.3m height limit that applied to buildings in Melbourne from 1916 to the 1960's was respected in the approved building on the right completed in 1992, alongside a significant building from 1928. The new development was combined with the old in such a way as to allow the new façade to be read separately within a continuous streetscape that keeps the old façade prominent.[21]

In addition to expanding concepts of heritage a landmark project was undertaken by the Australian Government in 1995 to develop an ethical methodology for cultural mapping that relates to all dimensions of culture, including its tangible and intangible elements. Published as *Mapping Culture – A Guide to Cultural and Economic Development in Communities*[22] the Guide is designed to help articulate cultural diversity and values through exploring culture in a comprehensive fashion. Positioned to redress the positivistic focus of the cultural heritage sector the technique seeks to encourage the sharing and exchange of culture and values – through the sharing of food, stories, values, and histories and places.[23] Revealing the diversity of cultural values and associations and creating new readings and meanings is achieved in partnership with communities, specialists, and custodians of knowledge. The methodology drew on the development of indigenous cultural rights in Australia, including land rights, which had strengthened considerations of attachments to land and their social value across Australia. (In 1992 the High Court of Australia in its 'Mabo' decision recognised native title to land and a process to enforce this.) This period in Australian history also illustrates the importance of non-mainstream cultures in influencing dominant cultural concepts, especially in the area of variations in cultural meaning. The 'Mabo' decision had the effect of deepening white perceptions of the holistic nature of culture, including community attachments to place, the importance of memory and stories and the indissoluble links between

---

20 P. Marquis-Kyle and M. Walker, *The Illustrated Burra Charter: Good Practice for Heritage Places* (Burwood: Australia ICOMOS, 2004).

21 Ibid., p. 38.

22 I. Clark, J. Sutherland and G. Young, *Mapping Culture – A Guide for Cultural and Economic Development in Communities* (Canberra: Commonwealth Department of Communications and the Arts, 1995).

23 G. Young, 'Cultural Mapping: Capturing Social Value, Challenging Silence', *Assessing Social Values: Communities and Experts* (Canberra: Australian Heritage Commission, 1994); and G. Young, 'Cultural Mapping in a Global World', Keynote Speech, ASEAN Cultural Minister's Conference, Adelaide, 2003), viewed 18 November 2004. <http://.www.fbe. unsw>.

Aboriginal and white culture, as manifest for example in the pastoral industry and in urban sites associated with the Aboriginal movement for civil rights.

Expanding Culture – Expanding Planning

Culture has a complex and uneven history in terms of its integration in planning. To focus this account I introduce three key themes and describe the picture mainly in Australian terms. The three themes encompass the postmodern world and its search for social communication and meaning, minority culture and the qualities of the new post-industrial age. Each theme has a particular planning reality that can be separated out quite clearly. For example, postmodern theory in Australia has provided new categories and themes in the realm of strategic planning for marketing, tourism and interpretation purposes, and new phenomenological and hermeneutic tools sensitive enough to capture important community values. These have been introduced in the area of migrant culture and the spaces of home.[24] Planning for indigenous culture in the postcolonial context of Australia has also achieved some notable innovations especially in relation to understanding the importance of variations in cultural meaning. Again, in post-industrial planning for cityports such as Sydney planners has sought to integrate regional and local culture in the form of ways of life, history, heritage and cultural meanings and traditions, in regional renewal strategies that are more sustainable.

Postmodern culture

Postmodern social and cultural theory emphasises the rise and importance of values, difference and diversity in cultures. In an era such as this, planning has a critical role in recognising and responding to the diversity of all ways of life as they evolve and are made manifest. The intellectual's role as a cultural interpreter emerges in this context as primary, and the planner is a facilitator in the process with communities. The research and recording of culture has become a means for communities and stakeholders to give voice to inclusive aspects of culture that might otherwise remain outside the picture. Cultural mapping as developed in the Australian guide cited earlier, is a structured technique positioned to capitalise on the whole culture of a place. In the Guide it is described in the following way:

> Cultural mapping involves a community identifying and documenting local cultural resources. Through this research cultural elements are recorded – the tangibles like galleries, craft industries, distinctive landmarks, local events and industries, as well as the

---

24 S. Thompson, 'Home: A Window on the Human Face of Planning', in R. Freestone (ed.), *Spirited Cities, Urban Planning, Traffic and Environmental Management in the Nineties* (Annandale: The Federation Press, 1993); and H. Armstrong, 'Cultural Continuity in Multicultural Suburban Places', in K. Gibson and S. Watson (eds), *Metropolis Now* (Sydney: Pluto Press, 1994).

Figure 3.2   Digital technology facilitates the mapping of culture

Source: Website Cultural Map of the Australian Captial Territory (ACT), Australia

intangibles like memories, personal histories, attitudes and values. After researching the elements that make a community unique, cultural mapping involves initiating a range of community activities or projects, to record, conserve and use these elements ... the most fundamental goal of cultural mapping is to help communities recognise, celebrate, and support cultural diversity for economic, social and regional development.[25]

Cultural mapping, is in a position to give voice to cultures and communities that would otherwise be unnoticed, neglected, or suffering marginalisation within a wider metropolitan stage, or urban research and policy context. It also has a simpler dimension as in the Cultural Map of the Australian Capital Territory

---

25 Clark, Sutherland and Young, *Mapping Culture – A Guide for Cultural and Economic Development in Communities*, p. 1.

Figure 3.3   Sydney 'surfboards' satirising the city's iconography and beach culture. Sydney's Gay and Lesbian Mardi Gras, Australia.
Photo: Garrie Maguire.

(ACT)[26] developed using elements of the flexible cultural mapping methodology. A 'top-down' map prepared by the ACT Government, the map is a website with an information and tourism role and permits viewers' insights into the ACT in terms of click-on categories for the Territory's history, multicultural culture, Aboriginal culture, places, people, events and tours (http.www.culturalmap.act.gov.au/). A postmodern dimension to the map lies in the use of contemporary digital technology to facilitate access to culture.[27]

Minority culture

In a postcolonial world overlain by a North-South divide, the movement of people and the growth of migrant communities in large urban centres pose specific challenges for planning to come to terms with diversity and to promote equality and harmony in society. Such planning will be based on understanding and accommodating difference. Minorities possess a shared history of difference and a burden of disadvantage that may be addressed in its cultural specificity by planning. Migration has been a common and enduring thread in the experience of the twentieth

---

26 ACT Government, *A Cultural Map of the ACT*, (Canberra: ACT Government, 2001) <http.www.culturalmap.act.gov.au>.

27 Young, 'Cultural Mapping in a Global World'.

century and the twenty-first is in this respect no different. Cities such as Sydney have experienced waves of migration from Europe and Asia over the decades since the Second World War and the communities based on this migration have re-defined society under the policy of multiculturalism and cultural diversity. Gay and lesbian communities have also grown in strength and diversity in Sydney and kindred cities and have contributed to and been beneficiaries of programs and planning respectful of diversity and its awareness. Examples of so-called 'pink' planning in Sydney have reduced the levels of gay and lesbian harassment and vilification. This has occurred by limiting uses such as entertainment parlours that attract aggressive patrons to gay oriented areas, notably Oxford Street, the route of one of Australia's biggest events the annual Gay and Lesbian Mardi Gras.

At every level planning then may be closely involved in strengthening communities and in encouraging broader community recognition and acceptance of diversity. This diversity exists in values and ways of life whether these are reflected in different socials needs, concepts of heritage and community place, or of the nature and importance of domestic family space. To do this successfully planning must rely on up to date concepts of culture and techniques such as cultural mapping if it is to 'read' the community in subtle terms and to engage it as a partner. In these cases the gastronomy, music, writing, religion and politics of these communities are all cultural references that assist the planner in the work of interpretation, for culture is the basis of understanding difference.

In Australia indigenous communities have embraced cultural research including cultural mapping. Projects include the mapping of attachments to rural land or 'country' and urban places, the history of families and sustainable environmental and ecological practices. Bi-cultural history has emerged and is the basis of bi-cultural strategic planning for cultural heritage places integrating different cultural knowledges and techniques in a shared vision. A vision that draws on local cultural knowledge, respects custodianship and traditional indigenous law and presents the planning framework in terms and language accessible to both sides.

Post-industrial culture

Culture is now being strongly engaged at the strategic level in regional planning in developed countries, especially in promoting the claims of major cities as centres for investment and visitation. On another level culture is also being promoted by the WCC and the World Bank as an integral part of any development project or program in the developing world. These two approaches have culture in common as both mobilise culture as the key to environmental sustainability and social and economic success.

For example, with the decline of waterfront areas in port cities globally in the 1980s culture came to be successfully mobilised in a number of key cities to provide the strategic basis for re-invention based on regional differentiation and to satisfy the authenticity demanded by local communities. In Barcelona, Spain a combination of museums devoted to the urban, artistic, design, maritime and political history of

Figure 3.4   A Harbour island's heritage of naval and maritime industry.
Cockatoo Island, Sydney Harbour (from the air), Australia.
Photo: Sydney Harbour Federation Trust.

Figure 3.5   A community festival celebrating an urban heartland.
Cockatoo Island Festival, 2005, Sydney Harbour, Australia.
Photo: Sydney Harbour Federation Trust.

the city, refurbished waterfront areas, promenades and attractions built on the city's cultural strengths and improved its image. The effect was a practical and perceptual renaissance. In Venice, Italy the city's historic arsenal area, neglected for most of the twentieth century, has been revitalised as the basis of high-tech centre for marine research and the development of technologies for environmental repair, together with major conference facilities. In Sydney, Australia the significant heritage of Harbour islands relating to convictism, ship-building and naval industries, for example, has in many cases been conserved and adapted with new, compatible uses. At the famous Cockatoo Island with its convict buildings and historic dry-docks, these uses include film-making and various kind of entertainment and spectacle, such as the Cockatoo Island Festival. This approach builds on an earlier strategic planning vision that emphasises the indigenous and multicultural history of Sydney Harbour and its role as a migration gateway.[28]

These three cities take their place in a post-industrial world that thrives on culture as a source of productivity and wealth. Thus, the cultural strengths of cities and regions can promote urban creativity and ultimately a successful image for attracting visitors, mobile capital and symbolic workers. Peter Hall describes this process and its results in the following way:

> Rich, affluent, cultivated nations and cities sell their virtue, beauty, philosophy, their art and their theatre to the rest lf the world. From a manufacturing economy we pass to an informational economy, and from an informational economy to a cultural economy.[29]

A key aspect of the passage Hall describes is that all life and all culture become increasingly self-referential, and it is to the implications of this I now turn.

For Culturisation

In addition to superficial approaches and the commodification of culture in planning, other issues impact on the extent and quality of cultural incorporation. Culture is subtle and complex in nature, its concepts are fluid and abstract, and there is a lack of understanding of suitable techniques and approaches for accessing and incorporating detailed and qualitative cultural knowledge in planning. Yet, culture expresses like nothing else the connective in life.

How then is this connective integration to occur, with potential transformative possibilities for planning? For example, how may culture be defined and made graspable for the planner given culture's many forms and emergent realities? How may cultural theory and planning theory in their oppositional diversities be harnessed for planning? How is it possible to develop a workable system to increase

---

28 G. Young, 'Planning the Addition of Value to Sydney, Australia's World City of Culture and Ecology', Unpublished proposal to the Hon. Bob Carr, Premier of NSW (Sydney: 1997).

29 P. Hall, *Cities in Civilization* (London: Phoenix, 1998), p. 8.

the authentic integration of culture in planning with potential benefits to planning independent of a single theoretical or political perspective? And would such a system produce beneficial outcomes in planning regardless of the value conflicts of a postmodern world of cultural diversity? A research enterprise of this kind is a radical undertaking and raises key questions. However, I believe it is possible to propose such an approach and to answer these pressing questions in the affirmative, by defining a role for culture in both spatial and strategic planning that has the potential to transform societies and their economies. It should also have lessons for other planning, and research and development in related sectors, such as education, health, governance, and the creative industries.

Before sketching the contours of such an approach I propose to deal with a response that has emerged to the re-invigoration of planning under the banner of 're-enlightenment'.[30] This is one strand in the response to the marginalisation of planning in the philosophical and market context of a triumphant economic neo-liberalism. The reason for this is firstly, that such an approach re-introduces intellectual and planning history to the debate, and secondly, it can be used to highlight significant contrasts with my own concept and paradigm of culturisation that should promote a better understanding of both approaches.

Culturisation or re-enlightenment

Firstly, I acknowledge that both culturisation and re-enlightenment approaches adopt a positive view of the activity of planning, and share the perspective that the practice of planning is too venerable in its history of positive social and economic engagement to be left to contemporary political cooption or class-based culture and mindsets within the planning profession.[31]

However, in the face of this reality the approaches differ. Any attempt to re-modernise or re-enlighten planning by returning to the universal values of the eighteenth century Enlightenment and its 'unfinished' project,[32] will be poorly adapted to a postmodern age of culture. Culturisation is able, through its positioning to respond to today's 'informational mode of development (in which) the source of productivity lies in the quality of knowledge'.[33] Re-enlightenment, I fear, remains a curiously quixotic and indeed ahistorical undertaking in an age of twenty-first century diversity, that follows a century that gave birth to total war, genocide and postmodernism. Within the context of an expanding and hybrid culture, a return

---

30 B. Gleeson, 'Reflexive Modernisation: The Re-enlightenment of Planning?', *International Planning Studies*, 5/1 (2000): pp. 117–135.

31 B. Gleeson, 'The Contribution of Planning to Environment and Society', *Australian Planner*, 40/3 (2003): pp. 25–30, p. 30.

32 U. Beck, *The Reinvention of Politics: Rethinking Modernity in the Global Social Order,* trans. M. Ritter (Cambridge: Polity, 1997); and Gleeson, 'Reflexive Modernisation: The Re-enlightenment of Planning?'.

33 M. Castells, *The Informational City. Information Technology, Economic Restructuring, and the Urban-Regional Process* (Oxford: Blackwell, 1989), p. 10.

to Enlightenment values or the monolithic perspective of modernism is unlikely to assist planning to regain its spirit or virtue. Whatever enlightenment may come, and however incrementally, is more likely to be the product of a heightened and 'educated' empathy for culture, diversity and incommensurability. These are the qualities required of a subtle age of culture that depends on an imperative of cultural interpretation, as well as the complex parameters and ramifications of global ethics and global governance.

Further, the literature on urban global competitiveness suggests that factors such as openness and diversity are behind the success of one region or place as against another:

> Talent powers economic growth, and diversity and openness attract talent ... Building a vibrant technology-based region requires ... creating lifestyle options that attract talented people, and supporting diversity ... These attributes make a city a place where talented people from various backgrounds want to live and are able to pursue the kind of life they desire.[34]

Re-enlightenment is ill-adapted to maintaining the multiplicities of culture and the nuances of culture that writers such as Florida highlight.[35] Similarly, the new generation of planning techniques such as cultural mapping work across different epistemologies and depend on the cross-fertilisation of collaborative and postmodern approaches and insights. This is a strength that seems to stretch beyond the ordered cosmos of the concepts of the re-enlightenment school.

An emphasis on culture and its strategic importance to planning, however, brings into the open a range of key questions in relation to the nature of culture, its distinctive operations and specific opportunities in the contemporary world. For example, what are the more authoritative and integrated concepts of culture the thesis seeks to mobilise and how is their transformative potential to be realised in planning? Additionally, what are the forms, scales and levels of planning under discussion and how do these diverse planning forms and modalities interrelate, in what are, after all, complex social and economic fields?

Positionality

It seems to me that the answer to these questions lies in a penetrating reflection of David Harvey's contained in his essay 'Social Justice, Postmodernism and the City'.[36] Harvey asks the question how might we now think about urban problems

---

34 R. Florida and G. Gates, 'Technology and Tolerance: the Importance of Diversity to High-Technology Growth', in T. Clark (ed.), *The City as an Entertainment Machine* (Amsterdam: Elsevier JAI, 2004), p. 214.

35 R. Florida, *Cities and the Creative Class* (New York: Routledge, 2004).

36 D. Harvey, 'Social Justice, Postmodernism and the City', in A. Cuthbert (ed.), *Designing Cities. Critical readings in Urban Design* (Oxford: Blackwell, 2003).

and 'how by virtue of such thinking we can better position ourselves with respect to solutions'.[37] He concludes that:

> The question of positionality is ... fundamental to all debates about how to create infrastructures and urban environments for living and working in the twenty-first century.[38]

The point of view I adopt is that culture is the key to the problematic of positionality raised by Harvey. A planning positionality focussed on solutions must be grounded in culture. This same recognition has emerged globally. It is the formal position of key organs of global governance and of international NGOs. For example, the World Commission on Culture (WCC), the key adviser on culture to the United Nations Educational, Scientific, and Cultural Organisation (UNESCO), expresses the view that culture is the foundation of all strategic planning for development. The WCC argues that:

> When culture is understood as the basis of development ... the very notion of cultural policy has to be considerably broadened. Any policy for development must be profoundly sensitive to and inspired by culture itself ... [39]

The Commission's understanding of this approach – utilising culture as the inspiration for development policy – is broad in nature. What this indicates is that in a quite conscious fashion our age is beginning to define itself in terms of culture and that the location, articulation and integration of culture in social technologies is now a dominant concern. This phenomenon represents a trend that is as much a key to the renewal of planning, as it is important to effective practice in other sectors. The WCC nominates a broad spectrum of interrelated areas as beneficiaries of culture, including education, health, the environment and organisational development. This will occur through a more thoroughgoing and sensitive approach to culture, based on planning and research, and development. One of the themes of this thesis is the striking opportunity cultural positioning offers planning to increase integration within and between planning at multiple scales and in different forms.

I argue in conjunction with the views of the internationalist organisations that the integration of culture into practical planning is perhaps the most important challenge in existence for planning today, and into the foreseeable future. This potential spans the full spectrum of planning forms, scales and purposes and represents the gravitational centre for planning reform. The intellectual introduction of the whole of culture into planning considerations is the key to the new positionality posited by Harvey.

---

37 Ibid., p. 101.

38 Ibid., p. 101.

39 C. Gordon and S. Mundy, *European Perspectives on Cultural Policy* (Paris: UNESCO, 2001), p. 5.

Conclusion

For culture to speak like Vladimir Nabokov's powerful and all-engaging memory would mean its detailed and reflexive presence in all planning. However, there is no single cultural or planning theory available to mobilise this transformation, and indeed, no single theory would ever be likely to account for the complexity of postmodern society. Yet, I would argue that the perspective of culturisation can promote an integrated approach to the concept of culture and a similar integrated approach to research. Over time these approaches should assist in the better synthesis of culture in planning and serve to promote to the exploration and articulation of the relationships between planning, society and the environment. A position of integration would be based on a comprehensive concept of culture and a plural approach to theory. An integrated view of culture could supplant the often narrow, selective and non-inclusive concepts of culture that lead to blinkered planning. Culturisation should promote a position of pluralism in respect of theory would draw in culture through a multiplicity of conceptual and methodological approaches. The complexity of this task is daunting, however, successful cultural approaches to planning do exist, although they are geographically and historically scattered and exist mainly in the form of *ad hoc* or piecemeal innovations, and the work of numerous planning theoreticians from both the neo-modern and postmodern wings of planning theory, contain innumerable insights and relevant observations that are not followed up. Culturisation would promote the development of a general cultural approach, operational concepts and a cultural model for planning that is required to fill the current vacuum in day-to-day planning needs. Over time, culture would be exhorted to speak more fully of itself.

PART II
Images and Heritages

Chapter 4

Capital Cities and Culture: Evolution of Twentieth-century Capital City Planning

David L.A. Gordon

David Gordon Associate Professor, School of Urban and Regional Planning, Queens University (Canada). Recent publications include Planning Twentieth Century Capital Cities (Routledge, forthcoming 2006), Remaking Urban Waterfronts (Urban Land Institute 2004), Battery Park City: Politics and Planning on the New York Waterfront (Routledge, 1997) and numerous articles on plan implementation and Ottawa planning history.

The twentieth century was a growth period for planning capital cities. There were about 40 nation states with capital cities in 1900. When the French and British empires dissolved after World War II, it was common practice to prepare plans for the new capital cities. By 2000, there were over 200 nation-states, with 191 members of the United Nations.[1]

Not all capital city planning was caused by dismantling empires and creating new nation-states. Some nations granted more autonomy to sub-state regions, leading to expanded government and cultural facilities in major provincial capitals like Barcelona or Edinburgh or in smaller self-governing aboriginal populations such as Iqaluit, the capital of Canada's new Nunavut Territory.

These new capital cities were developed during a century when urban planning was expanding as an administrative activity.[2] This chapter explores the elements that make up these capital cities and the evolution of their planning in the twentieth century, with some emphasis on cultural facilities. The research is based upon a

---

1 *World almanac and encyclopedia* (New York: World Telegram, 1898). Arturo Almandoz (ed.), *Planning Latin America's Capital Cities 1850–1950* (London: Routledge, 2002).

2 Anthony Sutcliffe (ed.), *Metropolis 1890–1940* (London: Mansell, 1984). Gordon E. Cherry (ed.), *Shaping and Urban World* (London: Mansell, 1980). R. Freestone (ed.), *Urban planning in a changing world: the twentieth century Experience* (London: E&FN Spon, 2000). Peter Hall, *Cities of Tomorrow*, 3rd edition (Oxford, 2000). Peter Hall, *Cities in Civilization* (London: Weidenfeld and Nicolson, 1998). S.V. Ward, *Planning the Twentieth Century City: The Advanced Capitalist World* (New York: Wiley, 2002).

study of 16 capital cities conducted by an international team of scholars.[3] Canberra and Iqaluit are featured as case studies that bracket the twentieth century.

What is a Capital City?

We defined a capital city as the seat of government for a nation-state, province or supra-national organisation like the United Nations or the European Union.[4] A typological framework developed by Sir Peter Hall was used to select and classify the 16 cases, which included Berlin, Brasilia, Brussels, Canberra, Chandigarh, Dodoma, Helsinki, London, Moscow, New Delhi, New York, Ottawa, Paris, Rome, Tokyo and Washington.[5] The most important differences in planning could be seen when comparing the plans for newer political capitals (Brasilia, Canberra, Dodoma, Ottawa, New Delhi, Washington) and the more established metropolitan capitals (London, Paris, Tokyo).

Most capital cities contained unique elements associated with their status as a seat of government, such as the parliament, embassies, cultural facilities and appeal courts (Table 4.1). In the newer political capitals (Brasilia, Canberra, Dodoma), it was fairly simple to identify these unique components. But as political capital cities like New Delhi and Washington matured and expanded into major metropoli in the late twentieth century, the distinctions became less clear.

In comparison, capital city planning issues were almost buried in metropolitan or global cities. London, Tokyo and Paris are long-established capitals, with the infrastructure for the seat of government largely in place by 1900. Their twentieth century planning histories revolve around metropolitan planning issues and their pursuit of global dominance in finance and culture.

Capital cities have particular planning issues associated with the seat of government, official residences, embassy districts, memorials and monuments, cultural facilities, symbolic content, finance and political relations between the local and national governments. The unique physical and land use elements of capital cities are discussed below.

---

3   David L.A. Gordon (ed.), *Planning Twentieth-Century Capital Cities* (London: Routledge, forthcoming 2006).

4   We did not consider the more recent annual designation of a European 'Capital of Culture'. Several cities that are not seats of governments have used to improve this designation to improve their profile and attract cultural tourists.

5   Peter Hall, 'The Changing Role of Capital Cities', *Plan Canada*, 40/3 (2000): pp. 8–11. John H. Tayor, Jean G. Lengellé and Caroline Andrew, *Capital Cities: international perspectives* (Ottawa: Carleton University Press, 1993). Hall's framework was especially useful since we compared the planning histories of the capitals, rather than their urban histories. Hall's framework was first presented in Taylor, Lengellé and Andrew (1993) and revised for this research project as P. Hall (2000), *Plan Canada*.

Elements that are located in most capitals

The building that is the seat of government is usually the principal symbol of the capital city in a democracy. It is usually set in a prominent position in the centre of the plan for a capital city. During the twentieth century, these capitol buildings have become more separated from the fabric of the vernacular city, at first to give them pride of place and later for security concerns.[6]

The official residence of the head of state may rival the parliament building in grandeur and symbolic content, especially in autocratic states or ex-imperial capitals (New Delhi's Rashtrapati Bhavan). Most countries provide their head of state with a splendid building, which is used for entertaining, receiving diplomats and as a family residence. These residences are often located at a modest distance from the parliament building to emphasise the independence of the legislature from the head of state. In former monarchies, the central royal palace is often turned into a state art museum (the Louvre; the Hermitage), rather than have the new president inhabit the previous regal home.

Capital cities usually contain buildings that are repositories of national culture. In the earliest stages of the development of a modern capital, these facilities may be combined. For example, the national library and archives may be co-located in one building or there may be a single multi-purpose museum and gallery. These facilities begin to splinter into independent institutions after the nation matures, collections grow and the capital becomes more culturally sophisticated. In metropolitan capitals, the institutions may be scattered about the city according to historical accident (Paris, New York, Tokyo), but London planned a concentration of cultural institutions in Kensington in the late nineteenth century.[7]

In political capitals, cultural facilities may be grouped in order to present a critical mass of national institutions (the Mall in Washington; the Parliamentary Triangle in Canberra, Confederation Boulevard in Ottawa). This technique also reinforces cultural tourism planning, providing visitors with an organised route connecting the sites, in the manner first planned by Pope Sixtus V in Rome.[8]

Many capitals provide facilities for political and cultural visitors, starting with information and orientation centres that combine national narratives with typical tourism information that might otherwise be provided by the local government. Some capital agencies go further, with a broad mandate to interpret the capital to visitors through programmed events and tours (Canberra's National Capital Authority, Ottawa's National Capital Commission).

---

6   Lawrence Vale, *Architecture, Power and National Identity* (New Haven, CT: Yale University Press, 1993). Lawrence Vale, 'Mediated Monuments and National Identity', *The Journal of Architecture*, 4 (1999): pp. 391–407.

7   Michael Hebbert, *London: More By Fortune Than Design* (New York: John Wiley, 1998).

8   Spiro Kostof, *The Third Rome, 1870–1950: Traffic and Glory* (Berkeley CA: University Art Museum, 1973).

Table 4.1 Capital City Elements

Legend:
- ● Many/All
- ◗ Some/Shared
- ○ Few/None
- N Not Applicable

Capital City Types	Multi-Function			Global Capitals		Political Capitals						Former Capitals		P. r.	Super Capitals		
	Paris	Moscow	Helsinki	London	Tokyo	Washington	Canberra	Ottawa	Brasilia	New Delhi	Dodoma	Berlin	Rome	Chandigarh	Brussels	Brussels EU	NYC
Elements located in most capitals																	
Seat of Government	●	●	●	●	●	●	●	●	●	●	◗	●	●	●	●	◗	N
Key Government Dept. Headquarters	●	●	●	●	●	●	●	●	●	●	◗	●	●	●	●	◗	N
Official Residence of the Head of State	●		●	●	●	●	●	●	●	●	●	●	●	●	●	N	N
Official Residence of the Chief Political Leader	●	◗	●	●	●	N	●	●	N	N	○	○	○	N	●	N	N
Embassies, Legations, Consular Offices	●	◗	●	●	●	●	●	●	●	●	○	●	●	◗	●	●	◗
Final Court of Appeal	●	◗	●	●	●	●	●	●	●	●	●	○	●	◗	●	N	N
Places for National Celebrations	●	●	●	●	●	●	●	●	●	●	◗	○	●	◗	●	N	●
Repositories of National Culture																	
• A National Library	●	●	●	●	●	●	●	●		●	○	○	●	◗	●	N	N
• Archives	●	●	●	●	●	●	●	●	◗	◗	○	○	●	◗	●	N	N
• State Art Museum	●	●	●	●	●	●	◗	◗	◗	●	○	●	●	○	●	N	◗
• National Cultural Museums	●	●	●	●	●	●	●	●	◗	◗	○	◗	◗	◗	●	N	◗
Political and Cultural Visitors Facilities	●	●	●	●	●	●	●	◗	◗	◗	○	◗	◗	○	●	●	●
Memorials and Monuments	●	●	●	●	●	●	●	●	●	●	○	●	●	◗	●	N	●

Elements Usually in National Capitals

Government Dept. Offices

National Bank

National Performance Venues

• State Theatre

• National Opera House

• State Concert Hall

• National Ballet Theatre

• National Sports Stadia

National Place of Worship

NGO Headquarters

Elements Sometimes Found in a Capital

Research Institutes

Major University

Major Religion Headquarters

City Hall and Mayors Residence

Elements NOT Usually Found in a Capital

Private Corporation Headquarters

Stock Exchange

Large Industrial Complexes

Urban Elements Found in Capitals

Parks (capital grander)

Transportation Gateways

Residential Areas

Retail and Shopping Facilities

National events require streets that are designed for grand parades (Paris' Champs Elysees, New Delhi's' Rajpath, London's Pall Mall), and places of assembly for national celebrations, demonstrations and spectacles (Moscow's Red Square, Beijing's Tianamen Square, Paris' Place de la Concorde). These ceremonial routes and places are often adorned with memorials and monuments to national leaders and significant national achievements. One of these places is often planned as the nation's principal stage for national remembrance ceremonies, which may be different from the celebration sites to allow an atmosphere of solemnity and mourning, particularly for the casualties of war.[9] If a place is particularly successful in capturing symbolic content, like Washington's Mall, it may begin to get crowded with memorials, requiring delicate political planning for a public art programme.[10]

Embassies, legations, consular offices and ambassadorial residences are part of every national capital. Early in the twentieth century, embassies were often in grand houses in the best residential districts, like Belgravia in London. With the dramatic increase in the number of nation-states over the twentieth century and expansion of the size of delegations, the older residential districts began to be crowded with buildings that were out of scale and traffic generators, such as the US Embassy on London's Grosvenor Square. The larger countries began to split their facilities into a consular office building (chancery) close to the legislative district and an ambassador's residence in a posh neighbourhood.[11] The chancery may also function as a local cultural showcase, with galleries and performance spaces for visiting artists.[12]

The headquarters of the key government departments are almost always located in the same city as the seat of government. Some decentralisation of back offices began in the mid-twentieth century, but the top staff and ministers typically want to be close to the key politicians. The seat of the final court of appeal and headquarters of the national judiciary are also usually found in the capital city. To emphasise their

---

9   David L.A. Gordon and Brian Osborne, 'Constructing National Identity: Confederation Square and the National War Memorial in Canada's Capital, 1900–2000', *Journal of Historical Geography*, (Forthcoming).

10   US National Capital Planning Commission *Memorials and Monuments Master Plan – Draft* (Washington DC: NCPC, 2000). US National Capital Planning Commission, 'Foreign Missions and International Organizations' in *Comprehensive Plan for the National Capital: Federal Elements – Draft* (Washington DC: NCPC, 2004), section 11.

11   Elizabeth Wiener, 'Report recommends new areas for embassies', *The Northwest Current* (Washington DC), April 14, 2004, p. 3. The receiving neighbourhoods may begin to object to the loss of historic fabric and the invasion of foreigners with diplomatic immunity to local traffic and parking laws, as in the Turtle Bay district around the United Nations General Assembly in New York, or the Kalorama neighbourhood in Washington.

12   Isabelle Gournay and Jane C. Loeffler, 'A Tale of Two Embassies: Ottawa and Washington', *Journal of the Society of Architectural Historians*, 61 (2002): pp. 480–507. Wize, Michael Z., 'Embassy Row receives an exotic face lift: Countries are commissioning high-profile architects to create diplomatic missions that reflect their cultural heritage', *Los Angeles Times*, April 18, 2004, p. 18.

independence, they are usually located in separate buildings at some distance from the legislature, or even in separate cities.[13]

Elements that are usually found in a national capital, but sometimes located in other cities

National cultural and sports performance venues are sometimes located in a capital city to showcase the host nation's culture to state visitors and serve the capital's population. However, performance facilities are more likely to be spread among many cities than national libraries or museums. A metropolitan capital may host the nation's dominant cultural facilities, such as Paris' Opera Garnier or Stade de France. These metropolitan areas often have large and wealthy audiences that may be able to support a variety of performing arts and special-purpose facilities, sometimes built by the private sector (London's West End; Manhattan's Broadway) or philanthropic donations. They may include a state theatre, national opera house, orchestral music hall, national ballet theatre or sports stadia. Paris continues to plan g*rand projects* in this field to enhance its status as a world cultural capital.

At the other extreme, the newer political capitals usually suffer in cultural comparisons with larger metropolitan areas, often enduring decades with temporary facilities. Embarrassment about state visits often leads to construction of a multi-purpose performance venue, such as Washington's Kennedy Center or Ottawa's National Arts Centre. More special purpose facilities may develop over time, after growth of tourism and the capital's population, but the performing arts facilities in capitals seem to evolve more slowly than national art galleries and museums.

Many capitals include a national place of worship – a grand mosque or national cathedral that is the site of state celebrations, coronations and funeral services. The location of these facilities in older capitals often depends upon historical circumstances and on national policies on the separation of religion and state. However, religious institutions are past masters at reserving dominant sites. The role of religious facilities in twentieth century capitals was often intensely controversial. For example, Stalin ordered the demolition of the prominent Kazan cathedral in Moscow's Red Square. New, officially planned and designated national places of worship are not common in modern multicultural nation states composed of multiple religions and ethnic groups. Mainly Catholic Brazil has a centrally located national cathedral in Brasilia, but Canberra, New Delhi, Ottawa, Dodoma and Chandigarh have no official religious facilities built as part of their capital city plans.

Although the United States was founded as a non-sectarian state, Washington has a 'National Cathedral', completed in the twentieth century. L'Enfant reserved a site close to the Capitol for a cathedral,[14] but in the 1890s the Episcopal Church secured

---

13 Carola Hein, 'Luxembourg, Judicial Capital of Europe', in *The Capital of Europe: Architecture and Urban Planning for the European Union* (Westport: Praeger, 2004).

14 Frederick Gutheim, *Worthy of a Nation* (Washington: Smithsonian Institution Press, 1977).

a presidential charter and bought the top of Mount St. Alban, the most commanding hilltop in the District of Columbia. The cathedral was built with private donations from 1907 to 1990, and is open for services to people of all faiths. Its tower is the tallest point in D.C.[15] Thus, while the Capitol dominates the political panorama of the Mall, the cathedral holds the site closest to heaven.

The capital city is usually the headquarters for national non-governmental organisations, cultural alliances, trade associations and international agencies. These agencies locate in the capital to lobby the legislators and civil servants to advance their members' interests.

The central bank, or national reserve bank that serves commercial banks, is usually located in the national capital.[16] This bank is typically closely concerned with national monetary policy, and usually locates near the offices of the finance ministry. But changes in communications technology have allowed many other governmental offices to be decentralised. Research laboratories, information processing centres and other facilities, which do not require direct access to the public or the legislators, were among the first facilities to move away from valuable inner city real estate.

Elements that are often located in another city, but sometimes found in a capital

Universities and research institutes do not require close access to the legislature for their normal activities and may actually benefit from some distance from political influence. Although most capital cities now host universities, post-secondary institutions tend to be distributed widely within developed nations. While some metropolitan capitals are also home to their nation's premier universities (Paris, Tokyo, Helsinki), these are not their country's only institutions of higher learning. Indeed, there was some tradition of establishing European universities at some distance from metropolitan areas (Bologna, Oxford, Cambridge, Leuven). In the nineteenth and twentieth century, North American public colleges were often deliberately planned for sites away from the state/provincial capital to spread the benefits of public investment to the hinterland.

Twentieth century capital city plans typically incorporated a complete range of educational institutions. But they rarely called for nationally funded universities that might play a dominant role in research and education in the same manner that a national gallery might for collecting visual art. Canberra was perhaps the exception, with Griffin's plan providing the site for a national university, which was initially a unique combination of graduate studies and research funded by the national

---

15 *Washington National Cathedral: A National House of Prayer for all people,* www.cathedral.org, Accessed August 1, 2003.

16 The United States is a prominent exception with the Federal Reserve Bank decentralised to several regions, with the New York bank the most important.

government.[17] Similarly, scientific research institutes often seem to maintain some distance from the legislators, either in suburban locations or other communities.[18]

The headquarters of major religions are rarely in capital cities, even if a national place of worship is maintained. The practice varies depending on history and separation of religion and state.

Finally, the newer political capitals are often separate from the commercial centres, which attract the headquarters of major private corporations (Mumbai, Toronto, Sydney, New York, Frankfurt). The country's major bourse or stock exchange is usually located in these commercial centres. The need to present a good image to state visitors and tourists often leads to some resistance to locating large industrial or manufacturing complexes in political capitals. Metropolitan capital cities often have plenty of manufacturing, but the industries are often located at a distance from the capitol complex and cultural circuit.[19]

Elements of major urban areas that are also found in national capitals

Most cities have parks but capitals often have a larger supply of open space to provide impressive settings for national institutions, cultural festivals, monuments and other symbolic content. The plans for the political capitals all recommended extensive parks systems and built them over the long term. The national governments acquired the land for the plans early because they had access to the relatively small sums of cash required to pay agricultural prices and the legal power to expropriate holdouts. These were crucial advantages when compared to non-capital cities.[20]

---

17 S.G. Foster and M.M. Varghese, *The making of the Australian National University:1946–96* (Sydney Australia: Allen and Unwin, 1966). Roger Pegrum, *The Bush Capital: How Australia chose Canberra as its federal city* (Sydney: Hale and Iremonger Pty Ltd, 1983). K.F. Fischer, *Canberra: Myths and Models, Forces at work in the formation of the Australian capital* (Hamburg: Institute of Asian Affairs, 1984).

18 For example, Canada's National Research Council in suburban Ottawa or the U.S. Brookhaven National (U.S.) Laboratories on Long Island.

19 Ravi Kalia, *Chandigarh: the making of an Indian city* (New Delhi: Oxford University Press, 1999), pp. 157–59. Chandigarh is an exception to this tendency. It has attracted modern corporation headquarters and industrial employers because it has good urban infrastructure, a clean environment and many services, especially in comparison to older Punjabi cities. See Kalia (1999) pp. 157–159.

20 David Schuyler, *The New Urban Landscape: the redefinition of city form in the nineteenth-century America* (Baltimore: Johns Hopkins University Press, 1986). Charles E. Beveridge and P. Rocheleau, *Frederick Law Olmsted: Designing the American Landscape* (New York: Universe, 1988). Charles E. Beveridge and Carolyn F. Hoffman, *Olmsted, Frederick Law, Writings on public parks, parkways and park systems* (Baltimore MD: Johns Hopkins University Press, 1997). Witold Rybczynski, *A Clearing in the Distance: Frederick Law Olmsted and America in the 19th Century* (New York: Scribner, 1999).

Planning for transportation systems in capitals is slightly different from other cities. The national governments may fund grander facilities as 'national gateways' or ceremonial routes. The key ceremonial entrance moved from the train station to the airport in the last half of the century. Some care is taken to manage the first impressions of visiting heads of state, or important international bureaucrats. Terminal facilities are upgraded (Rome, Paris, Brussels) and parkways may be built from the airport to the capitol (Ottawa). Of course these facilities also improve the first experience of national and cultural tourists, which benefits the local economy. The capitals were among the first cities to plan transportation circuits for cultural tourists.[21]

Canberra – An Early Twentieth-century Capital City

In 1912 Australia commissioned a controversial international design competition for a plan for its new capital.[22] Walter Burley Griffin's Canberra plan (Figure 4.1) is now recognised as a classic landscape design, placing a group of government buildings on a slope overlooking an artificial lake. The plan also included a university, and a 'Recreation Group' (Figure 4.2) north of the lake, including a stadium, theatre, opera house, 'Galleries for the graphic and plastic arts; the Museums for natural history and archaeology; the Zoological Gardens, and the Baths and Gymnasia ...'[23]

Griffin was a landscape architect from Chicago who had practised in Frank Lloyd Wright's office and was familiar with Daniel Burnham's City Beautiful plans. He immigrated to Australia with his wife, the architect Marion Mahony Griffin, who had helped prepare the winning design.[24] Although the prime minister promised that '... this city will be a seat of learning as well as of politics. And it will be home of art...' at the groundbreaking,[25] the Griffins had little success implementing the plan.

At first, Canberra suffered from under-funding and neglect at the hands of its federal sponsors. Nobody wanted to live there: legislators, journalists, cabinet members, and even Governors-General fled for the more cultured confines of

---

21 Thomas Hall, *Planning Europe's Capital Cities* (London: Spon, 1997), pp. 22–26. Dennis R. Judd and Susan S. Fainstein, *The Tourist City* (New Haven: Yale University Press, 1999).

22 John W. Reps, *Canberra 1912: Plans and Planners of the Australian Capital Competition* (Melbourne: Melbourne University Press, 1997), Chapter 3.

23 W.B. Griffen, *The Federal Capital: Report Explanatory of the Preliminary General Plan* (Melbourne: Commonwealth of Australia, Department of Home Affairs, October 1913), p. 46.

24 Peter Harrison, *Walter Burley Griffin: Landscape Architect* (Canberra: National Library of Australia, 1995, R. Freestone (ed.)).

25 A. Fisher, 'Speech at the ceremony to lay the foundation stones for Canberra' (March 12, 1913). quoted in Davies, E.V., J. Hoffman and B.J. Price, *A History of Australia's Capital, Canberra* (Canberra: ACT Ministry for Health, Education and the Arts, 1990), p. 46.

Melbourne or Sydney as soon as the legislature rose. The Australian legislature and civil service were 'temporarily' housed in Melbourne, and showed little inclination to fund or relocate to Canberra. Provisional parliament buildings were erected in 1927 on the site suggested by Griffin, but most departments did not move until the 1960s. Cultural facilities were also limited until a small theatre complex opened in the Civic centre in 1965.

Canberra had serious implementation problems during the middle of the century[26] but development began to accelerate after World War II. Prime minister Robert Menzies championed the growth of Canberra, appointing a powerful independent implementation agency, the National Capital Development Commission (NCDC).[27] Since there were few urban designers in Australia at the time, Britain's pre-eminent consultant, Sir William Holford advised the NCDC in the 1950s. He encouraged the NCDC to implement Griffin's plan, but with the cultural facilities and new parliament located on the south edge of the new lake.

Canberra grew from a scattered town of 36,000 in 1950 to Australia's largest inland city, with 325,000 inhabitants in 2001. Griffin's original plan, superbly fitted into a bowl of hills, was built out in the 1960s. As Holford suggested, Lake Burley Griffin became a focus for new institutional facilities in including the National Library (1968), the High Court (1980), the National Gallery (1982), National Science and Technology Centre (1988) and a visitor centre.[28] After considerable controversy, the splendid new Parliament House (1988), designed by Aldo Giurgola, was built into the summit of Capital Hill (Figure 4.3) rather than the lakeside site suggested by Holford. The NCDC was dismantled and replaced in 1989 by a smaller National Capital Authority. This new agency concentrates on the elements of Canberra that are directly related to its symbolic position as the national capital. The NCA developed additional public art and cultural facilities in the Parliamentary Triangle, including Reconciliation Place (2002), which celebrates the country's relationship with its aboriginal peoples.

Iqaluit – A Late Twentieth-century Capital City

Iqaluit (pop. 6000) is one of the world's smallest, and newest, capital cities. In 1999, it was designated the seat of government for Nunavut, the new territory of the Inuit peoples in Canada's eastern artic region. The site had been an Inuit fishing camp in the past, but permanent settlement began in World War II with a United States Air Force base and adjacent town named Frobisher Bay. After the war, Frobisher Bay became the main administrative and supply centre for the eastern artic. The

26 David L.A. Gordon, Ottawa-Hull And Canberra: Implementation of Capital City Plans. *Canadian Journal of Urban Research*, 13/2 (2002): pp. 1–16.

27 John Overall, *Canberra Yesterday, Today and Tomorrow: a personal memoir* (Canberra: The Federal Capital Press of Australia Pty Limited, 1995).

28 The Australian National Sports Centre (1976) required a large site and was constructed in the Bruce suburb.

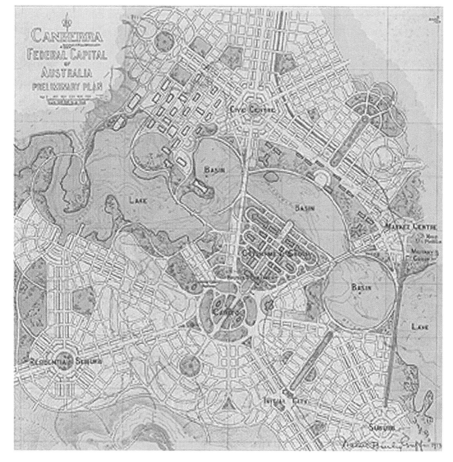

Figure 4.1   Canberra Federal Capital of Australia preliminary plan, 1914 by W.B. Griffin. The 'Recreational Group' is on the north shore of the central basin.

Source: National Library of Australia, Map G8984.C3S1 Gri 1913.

Canadian and Northwest Territories' governments both chose the town as their regional headquarters. Most residents were Canadians of European ancestry, but a few Inuit built a small village known as Ikaluit ('place of many fish') south of the base.

The early planning ideas for Frobisher Bay came from southern architects and planners who had limited experience with northern culture or environmental conditions. In the late 1950s, consultants to the federal government suggested that Frobisher Bay be redeveloped as a 'new town' in the Modern style and some

Figure 4.2 Organisational structure of the Canberra 'Recreational Group' proposed by Griffin. The cultural facilities were actually built on the south side of the lake.

Source: W.B. Griffin, *The Federal Capital: Report Explanatory of the Preliminary General Plan* (Melbourne: Commonwealth of Australia, Department of Home Affairs, October 1913) p. 8.

designs from the 1960s looked like spaceships dropped on the arctic tundra.[29] A small megastructure was built at the top of Astro Hill, with high-rise apartments, a hotel, offices, shops, a swimming pool and movie theatre to serve southern visitors and temporary workers. However, employment opportunities in the town began to attract more aboriginal residents and a more traditional arctic settlement with scattered single homes and commercial buildings grew up around the former military base. Plans from this era simply struggled to keep up with growth under difficult environmental conditions.[30]

---

29 Frobisher Bay Consultants, *Report for a Proposed New Town: Frobisher Bay, Northwest Territories* (Ottawa: Canada Department of Public Works, 1960).

30 J.L. Richards, 'Iqaluit: The Next Thirty Years', *Plan Canada*, 48/3 (2000): p. 37.

Figure 4.3   Aerial view of the Australian parliament with cultural facilities on the shore of Lake Burley Griffin in the middle portion of the photograph.

Source: Australian National Capital Authority, courtesy of Christopher Vernon.

The town was re-named Iqaluit in 1987 and by the late 1990s, it had acquired the elements of a provincial capital city – government offices, the Unikkaarvik Visitor's Centre, the Nunatta Sunakkutaangit Museum (Figure 4.4), an appeals court, and a small cathedral. These functions were housed in small buildings scattered across the core of the settlement.

The first plans adopted after Iqaluit became Nunavut's capital began to show some shifts in emphasis. The Nunavut government quickly built a legislature building on the main downtown street (Figure 4.5), and the surrounding area was designated as the Capital District in the 2003 Iqaluit General Plan. Government offices are intended to concentrate in this district, but the Legislature retained the option to build a 'new Legislative Assembly and other cultural facilities' on two sites with views of the water that are more traditionally representative of Inuit settlement patterns.[31] The 2003 plan also called for other capital city projects including a public square adjacent to the legislature to hold ceremonies and cultural events and a gateway from the airport to the core area to improve the image of the city for visitors.[32]

---

31 FoTenn Consultants, *City of Iqaluit General Plan* (City of Iqaluit, February 2003), p.42.

32 Ibid., pp. 9 and 15.

Figure 4.4   Nunatta Sunakkutaangit Museum, Iqaluit.
Source: Author's photograph, 2003.

The 2003 general plan incorporates several policies that reflect local conditions, including snowmobile trails, protection for waterfront fishing sheds, pedestrian routes and a cultural heritage designation that protected areas such as the traditional berry-picking patch.[33] More detailed policies in the 2004 *Core Area and Capital District Redevelopment Plan* (Figure 4.6) specifically address Inuit culture and recommend use of native materials (boulders) for landscape definition and public spaces throughout the town. The 'Big Artic Idea' for urban design is to represent all the communities by placing traditional carving throughout the district. The intention is to create a 'distinct Arctic character, that reflects the natural environment'.[34]

These latest plans also reflect some of the best practices from other winter cities.[35] The new core area plan recommends infill housing to increase the intensity of activity in the downtown area, while the design of the latest neighbourhood expansion calls for a more sustainable pattern of single dwellings, fitted more sensitively into

---

33 Ibid., p. 9.

34 FoTenn Consultants and Office for Urbanism, *Iqaluit: Core Area and Capital District Redevelopment Plan* (City of Iqaluit, October 2004), p. 83. This plan won the 2005 EDRA/ Places Award for Place Planning from the Environmental Design Research Association.

35 Norman Pressman, *Shaping Cities for Winter: Climatic Comfort and Sustainable Design* (Prince George BC: Winter Cities Association, 2004). Hough Stansbury Woodland, *Winter Cities Design Manual* (Sault Ste. Marie ON: City of Sault Ste. Marie and The Winter Cities Association, 1991). Pamela Sweet, 'Sustainable development in Northern Urban Area', *Plan Canada*, 44/4 (2004): pp. 41–43.

Figure 4.5   Nunavut Legislature in downtown Iqaluit
Source: FoTenn Consultants and Office for Urbanism, *Iqaluit: Core Area and Capital District Redevelopment Plan* (City of Iqaluit, October 2004).

the tundra landscape.[36] Southern firms prepared these recent plans with extensive consultation with local communities, since there are, as yet, no Inuit urban design consultants.

Diffusion of Capital City Planning Ideas

Stephen Ward's model of the diffusion of planning (see Table 4.2) differentiates between types of 'borrowing' (synthetic; selective; undiluted) and plans that were forms of 'imposition' (negotiated; contested; authoritarian).[37] In synthetic borrowing, the role of indigenous planners is very high and the external role is minimal. The

_____

36 SLB Consulting, FoTenn Consultants and Markbek, *A Sustainable Artic Subdivision* (Iqaluit: City of Iqaluit Planning and Lands Development Department, 2004).

37 Stephen V. Ward 'Re-examining the international diffusion of planning', in Robert Freestone (ed.) *Urban Planning in a Changing World* (New York: Routledge, 2000). Stephen V. Ward, *Planning the Twentieth Century City: The Advanced Capitalist World*. (New York: Wiley, 2002).

Figure 4.6 Iqaluit Core Area and Capital District Redevelopment Plan, 2004 (Annotations TBA)
A – Legislature; B – Canada offices; C – Astro Hill complex; D – museum; E – visitor center; F – proposed civic square; G – proposed gateway.

Source: FoTenn Consultants and Office for Urbanism, *Iqaluit: Core Area and Capital District Redevelopment Plan* (City of Iqaluit, October 2004).

1944 Greater London Plan, the 1956 Brasilia plan or the 1965 Paris SDAURP regional plan are examples where indigenous planners played dominant roles in creating and implementing innovative capital city plans. New Delhi's 1913 plan or the 1960 Iqualiut megastructure plan would be at the other extreme, as examples of authoritarian imposition of external planning traditions by an imperial or federal power.

Most twentieth century capital city planning could be classified as some form of undiluted or selective borrowing. Since most new capitals were created by the formation of a new nation-state, these cities had somewhat more choice of planning models than previous colonial settlements. Yet even relatively developed countries like Australia and Canada often lacked the specialised urban design and planning expertise to create a memorable capital city during the early twentieth century, so some borrowing was common, such as Canberra's 1912 Griffin plan. But there was wider variation in the implementation of the plans, and this is where the foreign models often ran into trouble, especially if they were far from their cultural homes. Although the smaller European countries and the British dominions engaged in

Table 4.2   Ward's Typology of Diffusion of Planning with Capital City Examples

Type	Indigenous Role	External Role	Typical Mechanisms	Level of Diffusion	Key Actors	Capital City Examples
Synthetic borrowing	Very high	Very low	Indigenous planning movements plus wide external contacts	Theory and practice	Indigenous	London 1944 Brasilia 1960 Paris 1965
Selective borrowing	High	Low	External contact with innovative planning traditions	Practice and some theory	Indigenous	Washington 1902 Brussels 1900s Helsinki 1910s Moscow 1930s NY (UN) 1940s Washington 1961 Canberra 1965+
Undiluted borrowing	Medium	Medium	Indigenous deference to innovative external planning traditions	Practice with little or no theory	External with some indigenous	Tokyo 1886 Brussels 1890s Canberra 1912 Ottawa-Hull 1915 Ottawa-Hull 1950 Chandigarh 1950 Tokyo 1950s Dodoma 1980s Iqaluit 1990s

Negotiated imposition	Low	High	Dependence on external planning tradition(s)	Practice	External with some indigenous	Dodoma 1970s
Contested imposition	Very low	Very high	High dependence on one external planning tradition	Practice	External	Brussels 1960s
Authoritarian imposition	None	Total	Total dependence on one external planning tradition	Practice	External	New Delhi 1913 Iqaluit 1960

selective or undiluted borrowing for their preparation of their plans, they retained control over implementation, smoothing the fit of the foreign models into the local context.

Canberra's twentieth-century planning history evolved from City Beautiful concepts imported via Griffin, to more functional designs for new towns and American land use/transportation planning. Freestone argues that more recent plans are more Australian and one can observe a transition from undiluted borrowing of Griffin's ideas to a more synthetic borrowing of planning concepts in recent initiatives.[38] In comparison, Iqaluit's planning history has barely begun. A portion of the megastructure imposed in the 1960 plan still dominates the skyline on Astro Hill. More recent plans acknowledge the settlement's role as the capital of Nunavut, but Inuit culture is still being filtered through southern planning and urban design consultants in a form of selective borrowing.

Conclusion

So, will capital city planning in the twenty-first century continue to be a substantial practice, or was this a twentieth century phenomenon? Certainly we are unlikely to see the explosive growth of the 45 years after World War II, when the number of nation-states tripled from 67 to 186.[39] In 1990, Anthony King declared that we have left this nation-forming phase and are well into the post-national phase where globalisation would be increasingly important. Certainly the global cities (London, Tokyo and New York) seem remarkably uninterested in urban planning in general, and national capital planning, in particular. And yet, the last decade of the twentieth century also saw perhaps twenty more new nation states, after the USSR, Yugoslavia and Czechoslovakia broke apart. There is no evidence that the separation of nations by ethnic, religious or language groups has ended, and the trend appears to affect both rich and poor nations. Even Canada, which spent the last years of the twentieth century with perhaps the world's highest quality of life,[40] faced separation of Québec in 1980 and 1995 referenda and created the new self-governing territory of Nunavut for its Inuit peoples.

Further, the impact of globalisation on other nation-states is not clear. Will they struggle harder to maintain national identity in the face of global pressures?[41] If so, the national capital may continue to be an important symbol and planned to encourage domestic and international cultural tourism. Or will these larger trading blocks, like

---

38 R. Freestone, 'The Americanization of Australian Planning', *Journal of Planning History*, 3/3 (2004): pp. 187–214.

39 A.D. King 'Cultural Hegemony and Capital Cities', in Taylor et al. (eds) *Capital cities/Les Capitales* (Ottawa: Carleton U.P, 1993) p. 252.

40 U.N., *Human Development Report* (New York: United Nations, 2002). As defined by the U.N. Human Development Index: UN (2002).

41 John Ralston Saul, *The Collapse of Globalism: And the Reinvention of the World* (Toronto: Penguin, 2005).

the European Union and the North American Free Trade Agreement (NAFTA) make it economically safer for regions like Catalonia and Québec to establish their own nation-states, and upgrade the status of their capital cities?[42]

Finally, the existing capital cities continue to grow and develop, since most of them have been the seat of government for less than 50 years. We noticed that the young cities add more cultural elements as they mature, and many continue to refine their plans as they grow. While the political role of capital cities may be in flux, the importance of cultural facilities in the planning of these cities seems likely to expand in the future.

Acknowledgements

The capital cities research benefited from advice from Eugenie Birch, Michael Hebbert, Javier Monclús, Ann Rudkin, Tony Sutcliffe and Larry Vale. The Canberra research has been assisted by comments from Rob Freestone, Christopher Vernon and Bruce Wright. Andrew Baigent, Pamela Sweet and travel support by the Canadian Capital Cities Organization assisted the Iqaluit research. An enthusiastic and talented team of research assistants contributed to the project, including Cheryl Dow, Laura Evangelista, Wesley Hayward, Ricky Hernden, John Kozuskanich, Kelly McNicol and Will Plexman. Canadian Social Sciences and Humanities Research Council grants and a Fulbright Senior Fellowship at the University of Pennsylvania financially supported the capitals research.

---

42 The Québec provincial government established the Commission de la capitale national du Québec in 1995 to guide the planning of its capital, Ville de Québec.

Chapter 5

The Power of Anticipation: Itinerant Images of Metropolitan Futures Buenos Aires 1900–1920[1]

Margarita Gutman

Margarita Gutman is an architect and urban historian, graduated at the University of Buenos Aires; Full Professor of Architecture and Urban History and holder of Chair at same university; Core Faculty at the Bachelor's Program and at the Graduate Program of International Affairs. Gutman was Scholar at Getty Research Institute and Woodrow Wilson International Center, Fellow at International Center for Advanced Studies, New York University, and Senior Fellow at the Vera List Center for Arts and Politics, The New School. Gutman is author and editor of 5 books; director of the exhibition 'Buenos Aires 1910: Memories of the World to Come' (1999/2000); director of BuenosAires2050 program, and member of the NewYork2050 initiative. She is preparing a book about anticipations of the urban future in Buenos Aires (1900-1920), analyzing graphic and textual representations of the future in the field of urban planning, media (popular magazines), caricatures, art, social and political proposals, and in the technological imagination.

Daily life in any society necessarily contains memories of the past, needs of the present, and expectations about the future. Images and ideas about future are tools that societies create to shape themselves. The construction of a sustainable and just future begins by imagining it. We need to take decisions and act today to get there tomorrow.

This chapter analyses images and ideas about the urban future produced and circulated in Buenos Aires around the turn of the twentieth century. At this time, Buenos Aires was experiencing unprecedented rapid urban growth triggered by

---

1   An earlier version of this chapter was presented at the Conference 'Locating the City', Antalya, Turkey, 2001, International Center for Advanced Studies, New York University and Bilkent University. A Spanish earlier version was presented at International Colloquium 'Buenos Aires / New York: Metropolitan Dialogues between South and North', Buenos Aires, May 2001, University of Buenos Aires and ICAS/NYU. Celina Baldasarre reviewed a large part of magazines in Buenos Aires, and Jasmine Mir in New York. Translation: Martha Morna Galvez and Alejandro Pareja.

economic, social and technological development. Anticipations and ideas about the future were pervading daily life and the subject of extensive and diverse public debates. The future was a driving force in society. The diverse ideas about the future composed a complex horizon of possibilities, containing available options at the time decisions were made, for both the private and public sphere.

To get to this horizon of available options of the urban future, this chapter analyses the urban anticipations published in Buenos Aires magazines, and compares them to ideas about the future contained in urban plans formulated by governmental and/or professional institutions. It contrasts two models of creation and representation of an urban future, originated in two distinct fields – that of the media and that of the professionals. This comparison allows for a closer understanding of the process of reproduction of the metropolis. It also shows the persistence of some images that, though not incorporated into any urban plan, have endured almost one hundred years, and had impacts evident in some parts of the city of today.

Secondly, this chapter analyses the City of New York as the implicit model of 'the city of the world to come' as published in Buenos Aires magazines through news about New York. Finally, this chapter discusses the graphic urban anticipations produced in New York for New York, and their relation with the anticipations published in Buenos Aires.

I. Magazines in Buenos Aires: The Empire of Technology

To analyse the representation of the urban future in the magazines published in Buenos Aires at the beginning of the twentieth century, we reviewed more than 2700 issues of a variety of magazines, the majority of which are large-circulation illustrated magazines published between 1901 and 1920.[2]

There is no doubt that the theme of the future is present in the magazines: in the 2700 examples reviewed, we identified 137 articles dealing explicitly with the future. They anticipated how the future would be and represented it through drawings and texts. Many articles were based, or claimed to be based, on scientific studies. Others discussed social or moral considerations, while others extrapolated on demographic or technological trends. They frequently imported visions of the future from other parts of the world. The majority celebrated, in earnest or in jest, the inventions and technical advancements that would radically transform daily life.

In spite of the great variety in the content of these anticipations of the future, a common denominator is persistently present: the city and urban life. Desired by the

---

2   Review of 2738 magazines, of which 2389 were published between 1901 and 1920. The magazines with the greatest number of issues reviewed were: *Caras y Caretas (CyC)* 608; *Fray Mocho* 468; *El Hogar* 340, *La Ilustracion Sud Americana (LISA)* 312, *PBT* 217; *La Vida Moderna (LVM)* 166; *El Sud Americano* 98; *Nosotros* 98; *El Sol* 81. Also reviewed: *Plus Ultra* 48, *Atlantida* 40, *Ars* 21, *Augusta* 19, *La Ilustracion Historica Argentina* 11, *Anales de la Sociedad Cientifica Argentina* 66, *Anales de la Biblioteca* 10, and a few of *Martin Fierro*, *Almanaque Sudamericano*, *Media Noche* and *Don Goyo*.

El mundo de mañana.

Figure 5.1   'El mundo de mañana' (Tomorrow's world)
Sources: 'The City of the World to Come' in Hudson Maxim, 'A vision of the future. Scientific and sociological prophesies,' *La Vida Moderna*, January 6, 1909, no. 27; 'The city of the world to come will be a single enormous construction...', *Caras y Caretas*, January 16, 1909, no. 537; 'A neighborhood of the city of the future', *Caras y Caretas*, April 9, 1910, no. 601; and Hudson Maxim, 'How the world of the future will be', *El Hogar*, February 2, 1917, no. 383.

Figure 5.2 'Buenos Aires in the year 2010'
Source: Arturo Eusevi, Buenos Aires in the year 2010, *PBT*, May 25, 1910, no. 287, special issue.

majority and feared by a few, the city is always on the map of the future in these magazines.

To illustrate the kind of urban future anticipated in the illustrated magazines, I chose an article that with some variations was appearing between 1909 and 1917 in various popular, large-circulation, illustrated magazines Those articles show images of the urban future that have persisted for at least a decade.

The most persistent image is the reproduction of one drawn by William Robinson Leigh in 1908, titled 'Visionary City', and inspired by the rapid growth of Manhattan.[3] It was included without attribution in the first Hudson Maxim article from *La Vida Moderna* in 1909. The illustration contains buildings evenly carved with windows, their facades devoid of ornamentation, with arches on elevated floors covering great

---

3   Joseph J. Corn and Brian Horrigan, *Yesterday's Tomorrow. Past Visions of the American Future* '(Baltimore and London: Johns Hopkins University Press, 1996), p. 34.

areas that appear to be vertical stations. There are multiple levels and types of bridges criss-crossing high above, intersecting with and connecting buildings. There is no street, no open sky – only vertical lines. There is not a complete line of vision, only a fragment, no bird's eye view, only a narrow vertical angle. Very few people are seen, just some pedestrians crossing the suspended walkways or traveling in cars via bridges hung over the urban abyss.

Between 1909 and 1917, that image was also published in other articles: in *CyC*'s January 16, 1909 issue, and 1910 *CyC* April issue. A variation of this image appears in the article published in 1910 in *PBT:* 'Buenos Aires in the year 2010', where there is also a reiteration of similar concepts.[4] The sketch artist, Arturo Eusevi, reinterprets for Buenos Aires the fundamental features of the 1909 'vertical city', with one difference: in the image of Buenos Aires in 2010 the sky is visible, the skyline is more open, and, above all, there are airplanes crossing above. Flying machines appeared in the image, and those apparatus 'heavier than the air' and their role in inter and intra urban transport are discussed. The airplane conquers the city's airspace. Later, in every image of the future of the city, airplanes would be present, as much as an image of anticipation 'would be' or as a caricature.

II. The Future of Buenos Aires Implied in the 'Lettered Plans' of the Time

During the first decade of the twentieth century, architects, engineers, and landscapers were involved with city problems from within diverse municipal offices, or from institutions like the *Sociedad Central de Arquitectos* (Central Society of Architects) and the *Universidad de Buenos Aires*. They argued, in public and in private, about the shape that the city of Buenos Aires should take in order to fulfill its new mission as the privileged symbol of a country in the midst of a population and economic boom. The republic was making its successful debut in the international commercial order, and Buenos Aires was comparable to the largest world capitals: it was eighth in population, and ranked high in wealth. Distinguished members of the political arena aspired to have Buenos Aires and Argentina play a leadership role in South America. The imminence of the centennial celebration of the 1810 *Revolucion de Mayo* and the diligent preparations for it, also brought about a new, critical look at the city and its future.

Many of these discussions were published in professional journals, such as the *Revista de Arquitectura* (Architecture Magazine), edited by the *Sociedad Central de Arquitectos*, and the annals of the *Sociedad Científica Argentina*. Excitement over adapting the old colonial city to meet the requirements of modern times was high, and plans were also presented from institutions such as neighborhood associations,

---

4   Enrique Vera y González, 'Buenos Aires in the year 2010', *PBT*, May 25, 1910 n287. In 1904 the author published *La Estrella del Sur. A traves del Porvenir* (The Star of the South. Through the World to Come) *Fabrica La Sin Bombo*, Buenos Aires, 1904. Second edition 1907, with illustrations by Buil, Herman Pujol, Rojas, Tusell, y Mendez Bringa.

**Figure 5.3   Intendencia Municipal Capital Federal, 'Modification project for
the current layout of the city of Buenos Aires, 1909'**
Source: Carlos M. Urien and Enzio Colombo, *La República Argentina en 1910*, (Buenos
Aires: Editorial Maucci Hermanos, 1910).

among others. The most discussed plans, however, were those developed within the
institutions mentioned above.[5]

The entire spectrum of the proposals, discussions, and disputes produced around
these plans, that we can call 'lettered', are marked by European influence.[6] From
France, and England came the cultural and scientific models and ideas that the country
had adopted for decades. Without any major discrepancies, Paris was the model of
urban excellence for Buenos Aires. Torcuato de Alvear, the first mayor, brought about

---

5   I comment on these plans in: Margarita Gutman and Jorge Enrique Hardoy, *Buenos
Aires, Historia del Area Metropolitana* (Madrid: Ed. Mapfre, 1992); Margarita Gutman,
'*Una ciudad en obra*' in Margarita Gutman (ed.), *Buenos Aires 1910:Memoria del Porvenir*
(Buenos Aires: Government of the City of Buenos Aires, FADU-UBA; IIED-AL, 1999).

6   Concept used by Angel Rama in *La Ciudad Letrada* (Montevideo, Uruguay: Comisión
Uruguaya por Fundación Internacional Angel Rama, 1984).

Figure 5.4   'The transformation of Buenos Aires'
Source: Villalobos, 'The transformation of Buenos Aires', *Caras y Caretas*, April 3, 1913, no. 757.

the initial major transformations towards modernizing the city, between 1880 and 1886. During his administration he received detailed reports from friends and special emissaries about the transformations taking place in the great European cities. Later, studies were carried out about the advancements Haussmann incorporated in Paris, and many French professionals were hired, such as Charles Thays, who was for many years the Director of Parks and Gardens of the Municipality of Buenos Aires. In 1906, mayor Carlos de Alvear solicited a plan for Buenos Aires from Joseph M. Bouvard, the Chief of Public Works of Paris. Together with a group of local professionals, Bouvard prepared the first plan for the entire city of Buenos Aires in 1909, but this plan was never sanctioned.

Nevertheless, the city had a series of urban regulations compiled in the *Reglamento General de Construcciones* (General Construction Decrees), which was approved in 1910. The shape of the city, through the layout of its streets, had been decided upon much earlier. In 1895, the municipality approved the first general map of the city, and in 1904 the official one was published showing the layout of all streets. The destiny of the city was decided upon by municipal civil servants, without any pretense of having established a special plan for the city.

The city discussed in the plans of 1900 and 1910 was horizontal and flat; the streets were laid out in squares, with low construction and only a few taller buildings downtown. The general impression was one of monotony, and a need for air and light in the narrow downtown streets designed by the tight colonial grid. The most pressing problem envisioned was downtown traffic, which in fact affected only a few blocks of the whole city.

Bouvard's plan presented a system of avenues and diagonals culminating visually in public buildings or plazas. They crossed the entire city as though they would be constructed just as they were drawn on the plan, reaching zones where there were not even open streets.

The engineer Benito Carrasco, developed another plan proposing the westward decentralisation of the city, towards the growth zones where there were still many vacant, low cost sites.[7] His plan proposed moving the public municipal buildings to less dense areas and contained a system of traffic circulation based also in a network of avenues and diagonals.

The themes most fervently discussed in publications were the design of diagonals and avenues: costs involved with municipal land acquisitions, the comparative advantage they would serve to alleviate traffic and the convenience or not of decentralizing the city. It seemed as if those discussions took place at the frontier of technical knowledge, demonstrated effectively by the incessant building of water and drainage networks, telephones and light, the construction of ports and granaries, bridges and subways during these years. The constant activity transformed the urban landscape of Buenos Aires into *'una ciudad en obra'* – the city as a work in progress.

The proposals of these 'lettered plans' concentrated on ornamentation, hygiene, and traffic. The avenues and diagonals, like the new grand parks and buildings, had fountains and monuments in strategically placed, visual locations. They allowed a clean and handsome context overcoming the monotony of the ancient colonial square grid. Paradoxically, that square grid was the feature recently decided upon for the city yet to be constructed.

III. 'The City of the World to Come' vs. 'The Lettered Plans'

In this section, we will briefly compare 'the city of the world to come' or 'the vertical city' – those anticipations published in mass, popular magazines – and 'the lettered plans' those elaborated by professionals, and published in specialised magazines.

Transportation: from the horizontal plane to three-dimensional space

The attention paid to the traffic problem in both types of anticipations shows the intense preoccupation with this urban issue. But traffic congestion in Buenos Aires occurred only in few blocks in downtown, while the rest of the city unfolded slowly, without any congestion problems whatsoever, for many decades to come.

---

7   Benito Carrasco, 'The city of the world to come', *CyC*, February 22, 1908, no. 490.

To resolve downtown traffic, the 'lettered plans' proposed solutions over a horizontal plane: a system of avenues and diagonals. In contrast, the 'vertical city' published in the magazines, details an unending paraphernalia of gadgets and solutions to resolve traffic congestion and to allow increased speed. These solutions focus on the utilisation of airspace and on the design of three-dimensional structures where transport was segregated by type and level. The air and underground had been conquered. In fact, the entire city was conceived as a new and marvelous container that would allow, above all, the circulation of people – particularly through the air: elevated trains and planes, dirigibles and hanging bridges, cables and cable cars.

Generic plans vs. plans for specific places

Another important difference between those anticipations is location: 'the vertical city' offers only generic solutions, not specific to designated places. In general, they have no location in the actual geography of the city. The 'lettered plans' obviously did designate many but not all locations.

No one believed that the solutions shown in magazines could ever be applied. But the letter plans were not implemented for those decades either. By 1925 another master plan for Buenos Aires, *Plan Noel*, was formulated, but not sanctioned. Later the *Oficina del Plan Regulador* (Plan Registry Office) was created, later producing the *Plan Regulador* of 1960 (The Regulating Plan). The city then had three million inhabitants, and Greater Buenos Aires had close to seven million.

The future: a guaranteed destiny

Both the future of the 'vertical city' and the one anticipated by the 'lettered plans' were assured; there was no doubt that they would come. Nor was there any doubt that the future would be an improvement. The verb tense used in magazine anticipations was the future indicative, and never the subjunctive. Even if the literate plans had difficulty getting approved, there were no doubts about the possibility that they could be made realities.

The future: a journey between Paris and New York

The 'lettered plans'' clearly proposed to emulate the model of Paris. The urban future in the magazines looked to the intensive application of advancements in science and technology to urban life and its challenges. It is no surprise that images from New York popped up everywhere.

Both fields, in their own way, placed Buenos Aires in the international realm. The 'lettered' field did so fundamentally, by incorporating the presumed universal values of urban beauty that were in reality particular to Paris. In the magazine anticipations, the 'universal' was directly related to the widespread application of technology and science onto the city; the city of New York was the universally recognised symbol of this achievement.

In each camp, the future seemed to be interchangeable with other cities. As an extreme simplification, we can say that the journey to the future of Buenos Aires extended from Paris to New York.

The three-dimensional structure vs. the two-dimensional plan

The 'city of the world to come' is a three-dimensional structure, extensive and complex: 'The city of the world to come will be an enormous construction, with a stupendous ridge of arcades, galleries, parks and gardens imposing themselves to an inconceivable height.'[8] It is not a discrete group of buildings placed on the plan of the traditional city, nor is it a two-dimensional map upon which public spaces and buildings or shelters for pedestrians, carriages, and throughways, are arranged.

Other public spaces, other streets, other skies

In 'the vertical city' there are no streets at ground level, or, at the very least, they do not serve the same purpose in the traditional city. The streets as we know them in the traditional city acquired other functions in the 'vertical city': they are secondary and supply routes, hidden and unseen. The old functions (movement, meetings, recreation, shopping, *flaneur*, aerobic exercise, etc.) change locations, or cease to exist. Nor does there exist open sky. In the majority of the drawings, fragments of urban structures appear in which neither streets nor sky are seen.

Fragments vs. totality

The 'city of the world to come' appears in the magazines as a fragment of a three-dimensional structure. Very rarely are their borders shown. Conversely, in the 'lettered plans,' the city is planned in its totality, from border to border. Even when the city was still mainly empty, it was perceived as a complete entity.

Flows vs. fixed space

The enormous importance that transport acquires in 'the city of the world to come' is expressed through a clear emphasis on images showing every form of movement of individuals – the flows.

Many fragments of skies vs. a single celestial cupola

The 'lettered' city of the future is covered by a single sky, a perfect celestial cupola. The 'vertical city', on the other hand, is covered with infinite fragments of skies crossing within its three-dimensional structure. In spite of its grandeur, and the weight of the structures shown in the drawings, the texts describe the city as an aerial

---

8   Hudson Maxim, 'How will the world be in the future?', *El Hogar*, February 2, 1917, no. 383.

city, light, 'resembling steel mesh', inundated by air and sun and basically healthy. Nevertheless, in the drawings, the city appears to be dark and almost uninhabited. Where are the people?

Light and electricity

Light is a critical factor in 'the city of the world to come': evident in countless nocturnal images of the city illuminated as if by day, crossed by speeding illuminated trains, surrounded by thousands of points of light from planes. In the 'lettered plan' and in the traditional city, artificial light seems to accompany the plan: illuminating fountains and cupolas, reliefs and cornices, and ornaments with wreaths of lights or lanterns. But they do not seem to constitute a key factor as in the 'city of the world to come'. The ebullience of the 'vertical city' by night presumes a certain continuous activity, at least in terms of transport. 'The city of the world to come' seems to be a 24-hour city. The traditional city sleeps by night.

And beauty? What beauty?

In the 'vertical city' the architecture concentrates almost completely on the expression of volume that is decidedly vertical. In barely a single case does ornamentation finish the cornices. More often, walls are marked with narrow orifices of neatly and evenly placed windows. The bridges, with their arches and steel wires, apparently respond to structural questions, without ornamental or decorative additions. Little remains of the Beaux Arts architecture in these illustrations: just a decorative fountain arranged as the center of a rotunda that organises a timid crosswalk between two aerial bridges. On the other hand, the 'lettered plans' are basically organised by aesthetic criteria.

Virtual infrastructure

In 'the city of the world to come' electricity is the source of the power to resolve every type of situation, but, fundamentally, communications would be wireless; 'telephone, *telarmonio* and television' would allow communication within the same city, and even 'with the Antipodes with complete ease'. The highest value of wireless communication is its potential for entertainment: bringing theatrical and musical spectacles performed in the great cities to remote hamlets or to ships in mid-transit. The network of infrastructure would be entirely virtual.

Environment and the suburbs

Nature, in the future as shown by the magazines, has been domesticated: 'when heat and fuel is inexpensive, the surface of the planet will be a pleasure',[9] a message repeated in one way or another in almost all the articles.

---

9 'The city of the world to come', *CyC*, April 9, 1910, no. 601.

Thanks to this provision of energy and to the means of transportation, people could live at great distances – to the suburbs enormously extending the space to be used for living areas. The 'lettered plan' assumes suburbs with a certain form, and incorporates domesticated nature in parks and plazas without any apparent tensions.

And the inhabitants, where are they?

'The city in the world to come' seems to be an image of 'all's well'; optimistic and secure. This future speaks of a solid confidence in progress and evolution. It supposes that technology will allow a better moral and social life. The technological advancements are understood as benefactors. But, who is benefiting? Few people are seen in the drawings, with little mention of their social or economic relations. Vignettes of daily life in the future appear only as caricatures, the majority of which deal with the impacts or echoes of scientific and technical inventions.

The end of privacy: 'a completely public life'

In one of the caricatures, there is an anticipation of the end of privacy and an increase in control: 'Completely public. A glance towards the future'.[10] The article's drawings appear to allude to a European city. Everything visible owes its existence to technological advancements, such as x-rays seeing through walls or a mechanism that registers thoughts – with a bell that sounds when it reads bad thoughts (close to a school the sound is deafening). 'Oh, nothing is private today; everything is public.' The wealthiest build underground to avoid being seen and to conserve their privacy. But, 'it's money thrown in the street. What would be the point of opposing the path of progress? Each thing must be entirely public, and will be'.

Transparency: to live in glass houses

In the end, for this future of daily life, crystal-clear transparency is imposed, such as a feature in 'To live in crystal houses'.[11] The article points out that 'The prophecies of Flammarion announced that in the 'world to come' humanity would live in transparent crystal buildings. They are starting to come true.' Both features of this vision of the future, the end of privacy and transparency, are discussed today.

---

10 'Completely public. A glance towards the future', unsigned, *El Sol*, February 9, 1908, no. 29.

11 'To live in a crystal house. The human habitat of the world to come', *El Hogar*, February 19, 1915, n. 281.

III. New York: Model of the 'City of the World to Come'

Regardless of the titles of the published drawings of 'the city of the world to come' or 'the vertical city', images of New York lurk behind all of them. Readers of these magazines could easily recognise the implicit reference to New York, even if it was not explicitly mentioned in the text. The extreme verticality of the structures, hanging bridges, submerged tunnels and elevated trains, passageways and hanging galleries linking large buildings, viaducts hidden under the streets spanning entire areas of the city, were all clearly inspired by the great building projects then underway in New York. These 'grandiose' works received abundant attention in Buenos Aires magazines, which clearly singled out New York as the – urban paradigm for the mass application of science and technology. Having this quality of technological frontier, associated with enormous commercial, financial, industrial, and port development, it can be said that New York was one of the models for an anticipated urban future for Buenos Aires at the beginning of the twentieth century.

But this review must be formulated in the context of the complex relationship that existed between the two countries at the time, when their dialogue was neither fluid nor easy. In fact, feelings about the United States were very contradictory: 'About no other country is general opinion so divided among our contemporaries', said in 1914 a New York contributor to the magazine *Nosotros*, referring to the opinion of Argentines about the United States.[12] Along these same lines, the magazines, while publishing countless articles expressing admiration for the 'colossal' advances being made in New York, also vented sharp criticism of North American foreign policy, especially towards Latin America.

In the Buenos Aires magazines, numerous articles contain information or commentaries about the United States, most of them about New York (infrastructure, buildings, avenues and urban design, statistics, social problems, organisation, everyday life, health, commerce, culture), and others about the United States in general (attitudes, culture, inventions, journalism, sports, fashion, women, humor, shows, movies). There are some articles detailing United States presence in Buenos Aires (advertisements for North American products – cars, office desks, shoes, gramophones, concrete – banks and insurance companies), and some containing political commentaries, as the critical political cartoon titled 'The Yankee Threat', starring 'Uncle Sam' bending over South America.[13]

New York as an image of progress

A large number of the articles about New York express admiration for the great building projects then underway, with photographs, drawings, and commentaries

---

12 Alejandro Jascalevich, 'Yankee Chronicles', *Nosotros*, June 1914, no. 62, p. 299.

13 'The Yankee Threat', *CyC*, Dec. 17, 1904, no. 324. Uncle Sam crouching over the American continent. Legend: 'Monroe Doctrine, […] no one lays hands here but me.'

Figure 5.5   'The Yankee Threat'
Source: 'The Yankee Threat', *Caras y Caretas*, December 17, 1904, no. 324

about bridges, tunnels, elevated trains, subways, and great terminals.[14] The cries of admiration reach their highest pitch in praising the great dimensions of Grand Central Terminal, whose progress was documented throughout several years in different magazines.[15]

In these magazines, New York was perceived as a place of constant change and reconstruction, a kaleidoscope of great contrasts and a model of technological progress.[16] But there were also warnings and criticisms: in 1919, a short article described New York as 'the kingdom of the incredible', did not see any aesthetic prodigy in skyscrapers, and denounced the lack of calm in an urban area razed by 'the neurasthenic noises of a great city.' Nevertheless, it admitted that 'no one can say that New York, seen at night from high up in the midst of the black calm of night, is not admirable, portentous.' The article also compared electrical light to the radiance of the spirit, 'illuminating the path to the future'.[17]

Lights, heights, and a Babel-like city

The city's lighting was enthusiastically celebrated: New York, with its network of 'white streets ruled by light', was, according to Buenos Aires magazines, 'the best illuminated city in the world'.[18] In 1915 *PBT* published 'American Nights'[19], which included nighttime photographs of Buenos Aires, Rio de Janeiro, and New York, with fully lighted avenues. These images, according to the author, proved that these two cities were on the same level as New York, leaving Europe behind and becoming part of the New World.[20] Lighting is used to allude to the future, and Buenos Aires's destiny is linked to America's destiny. The three cities are grouped under the same denomination, 'America', to invoke a future of infinite and shared progress. The article ends by looking ahead: 'What will America be in the future, what will be the destiny of this emerging force that surprises us today? The future that it augurs is

---

14 'The great new hanging bridge in New York', *CyC*, Oct. 27, 1906, no. 421; 'Tunnels under rivers', *CyC*, March 9, 1907, no. 440; and 'The great hanging bridge of New York', *El Sol*, Feb. 15, 1908, no. 28.

15 H. Bougeois, 'The Great Station in Central New York', *El Sol*, Year II, Oct. 30, 1907, no. 21 (alb. 34); 'The world's largest station', *LVM*, May 3, 1911, no. 212; and 'The world's largest station', *El Hogar*, Oct. 15, 1915, no. 315.

16 'The panorama of New York throughout four decades', *PBT*, August 26, 1916, no. 613; and 'The reconstruction of New York', *Fray Mocho*, Oct. 25, 1912, no. 26 .

17 'New York at night', *Plus Ultra*, May 1919, no. 37.

18 'New York on Christmas eve', *El Hogar*, Dec. 25, 1907; 'White streets: Empire of light in the cities of North America', in *Fray Mocho*, June 28, 1912, no. 19; and 'Electricity in New York', *Fray Mocho*, Nov. 30, 1920, no. 449.

19 'American Nights', *PBT*, April 10, 1915, no. 541.

20 The four photographs of Rio de Janeiro are identical to those published in 'The White Lights of Rio' and 'What Kind of New York is Buenos Aires?', *The World's Work*, 29 December 1914, pp. 192–208.

barely conceivable. The light that is now turning on illuminates the altar of infinite progress.'

The enormous height of the skyscrapers, those 'twenty-floor neighborhoods', attracted a number of photographers who produced a famous series of photographs about 'those who are not afraid of vertigo'. New Yorkers amazed the world by their audacity in 'defying the clouds in the sky'.[21] The great heights of its buildings and their photographic representation, its cosmopolitan and diverse population, along with other characteristics of New York suggested the obvious association with the Tower of Babel and Babylon. In 1920, an article entitled 'Modern Babylon'[22] included images of the Woolworth and Flatiron Buildings, the Brooklyn Bridge, and Broadway. The article called New York a 'supercity,' arguing that 'compared to New York, Paris is a toy Babylon.' New York is an 'unsettling and noisy whirlwind', where 'every large house is a Tower of Babel'. The article ends by wondering what New York will be like in the year 2000 and what influence it will have on the world.

Buenos Aires compares itself with New York

Buenos Aires had 'white' avenues illuminated at night too, as well as never before seen heights and fearless photographers. It is not surprising that the *'La Reina del Plata'* compared itself with pleasure and merriment to New York through the brave feats of its photographers. They too climbed up buildings under construction to photograph the city from above. In 1916's 'New York in Buenos Aires'[23] the city is not the only star of the photographs: the photographers themselves are stars, dressed in street clothes and hanging from the city's cornices: 'on the abyss's edge or *struggle for life*'[24] is the caption of a photograph taken from the top of the Post Office Palace, then under construction.

These comparisons were not new, since in 1890 Buenos Aires had already compared itself to New York. Evidence of this is an article published in *El Sud-Americano* that speaks of New York as 'the Commercial Capital of America, which in the near future will be surpassed by Buenos Aires'.[25]

'The other half' of New York, in Buenos Aires

In the same way that Buenos Aires magazines echoed New York's grandiosity, and its technological advances, they also focused on its poverty. News about 'the other half' highlighted the sharp contrast between luxury and misery in New York City.[26]

---

21 'New York, the city of skyscrapers', *LVM*, October 17, 1907, no. 27.

22 Roberto Lee, 'Modern Babylon', *Plus Ultra*, February 1920, no. 46. Article written in January 1920 in New York.

23 'New York in Buenos Aires', *Fray Mocho*, October 20, 1916, no. 234.

24 In English in the original.

25 'Lungs for our cities', *El Sud-Americano*, October 18, 1890, no. 61–62, p. 134.

26 Jacob Riis, *How the Other Half Lives. Studies among the Tenements of New York* (1890). Re-edited in New York by Dover Publications, 1971.

Figure 5.6   'New York in Buenos Aires'
Source: 'New York in Buenos Aires', *Fray Mocho*, Oct. 20, 1916, no. 234

In 'How poor New Yorkers live', we read: 'those who think that New York is the ideal city will undoubtedly be surprised by the misery that coexists in its streets with the most radiant manifestations of opulence and magnificence.'[27]

Tenant and trolley strikes were discussed in detail.[28] Growth in the immigrant population was also discussed, as well as the selection criteria applied to immigrants at Ellis Island.[29] New York crime and police news were also juicy topics.[30] New York fires and catastrophes were amply covered with commentaries and photographs. One of the most powerful photographs of the harshness of winter in New York shows ghostly ice stalactites hanging from structures twisted by a fire.[31]

IV. The Future of New York Inspires the Future in Buenos Aires

Observing the images of 'the city of the world to come' anticipated for Buenos Aires and circulated in the magazines, it is possible to suppose that they were not only inspired by the contemporary city of New York, but also by images created in New York itself to anticipate its own future.

This is evident in a series of images published both in New York and Buenos Aires. At times, the images published in Buenos Aires were obviously inspired by New York's, at others, they were direct reproductions. For instance, many of the features of the anticipation of 'the vertical city', analysed in section III, seem taken from or inspired by some of the drawings published in the series *King's Views of New York*, such as 'The cosmopolis of the future', by Harry M. Pettit, published in 1908, or 'The Towering Skyscrapers', published in 1911. These illustrated New York guides were published by Moses King, sold at low prices, and enjoyed wide circulation.[32] The same drawing by Harry M. Pettit was reproduced in a 1916 postcard, with the title 'Future New York City: The City of Skyscrapers'.[33] At least in the form of postcards and guides, this image seems to have circulated widely

27 'How poor New Yorkers live', *Fray Mocho*, February 9, 1917, no. 250.

28 'Effects of a trolley strike in a giant city', *PBT*, November 25, 1916, no. 636; and 'Tenant strike', *El Hogar*, September 7, 1917, no. 414.

29 'How they choose immigrants in the United States', *LVM*, March 8, 1911, no. 204.

30 'The Mafia in the United States', August 5, 1905, no. 357 (alb. 41); 'The child police of New York', *Fray Mocho*, Sept. 27, 1917, no. 283; 'The great silk robberies in New York', *Fray Mocho,* Sept. 10, 1918, no. 333; 'New York police', *Fray Mocho*, July 13, 1919, no. 429; and 'How women thieves operate in New York', *El Hogar*, December 18, 1914, no. 272.

31 'The cold in North America. View of a burnt house in New York. The water hurled by the fire fighters instantaneously turns into ice, *LISA*, April 30, 1903, no. 248, p. 126; 'The catastrophe in New Jersey', *Fray Mocho*, August 4, 1916, no. 223; and 'A great fire in New York', *LVM*, Year IV, September 21, 1910, no. 180.

32 Moses King (ed.), *King's Hand Book of New York City* (Boston, Mass.: Moses King, 1892).

33 Postcard in Rebecca Zurier, Robert W. Snydner and Virginia M. Mecklenburg, *Metropolitan Lives* (New York-London: National Museum of American Art, W.W. Norton and Company, 1995), p. 88.

Figure 5.7   **'The Towering Skyscrapers'**
Source: Cover of *King's Views of New York*, published by Moses King, 1911.

Figure 5.8   'Buenos Aires in the future'
Source: *El Hogar*, 1929.

Figure 5.9   'A vision of New York in the Future'
Source: Reproduced in Dietrich Neumann (ed), *Film Architecture. Set Designs from Metrópolis to Blade Runner* (Munich, London, New York: Prestel, 1999), p. 35.

Figure 5.9   Continued

Figure 5.10 **'New York City, as it will be in 1999'. Pictorial Forecast of the City as Approved by Andrew H. Green, H. H. Vreeland and John McDonald'.**

Source: *Supplement to The World*, New York, December 30, 1900. Reproduced in Judy Crichton, *America 1900*. *The Turning Point* (New York: Henry Holt and Company, 1998), pp. 260–261.

Figure 5.11   'Looking Forward-A.D. 2001: Broadway, New York, as it may
                   appear a hundred years hence, when modern inventions have
                   been carried to the highest point of development.'

Source: *Collier's Weekly*, Jan. 1901, reproduced in Crichton, p.2.

both in New York and abroad. Years later, in 1929, the same Pettit drawing would be reproduced in *El Hogar* under the title 'Buenos Aires in the future'. It was an operation of appropriating a future for Buenos Aires, similar to the one executed by *PBT* almost twenty years before, when it published in 1910 Arturo Eusevi's drawing 'Buenos Aires in the year 2010'.

Using similar forms, *Scientific American* published in 1913 an anticipation of the future New York drawn by an engineer titled 'A Vision of New York in the Future'. Now from a professional perspective, the vertical city was reproduced with subways, viaducts, and bridges.[34] In September 1913, that image was reproduced in Buenos Aires in the magazine *Caras y Caretas*, under the title 'The City of the Future. A daring solution to the traffic congestion problem.'[35] The epigraph only adds that it was made by a North American engineer, but does not say that it is a vision of the future made for New York. By re titling the image as 'The city of the future' the model – originated in and made for New York – was thus made available to any other city.

An urban future for New York in New York magazines

These images anticipating the future were linked for the most part to technological transformations of urban infrastructure, especially innovations in transport both within and between large cities. This is evident in two drawings published in late 1900 and early 1901 in *The World* and *Collier's*, respectively, which anticipated the look and lifestyle of New York City in 1999 and 2001.

A panoramic drawing titled 'New York City as it will be in 1999. Pictorial Forecast of the City as Approved by Andrew H. Green, H.H. Vreeland and John McDonald',[36] was published on December 30, 1900. It contains an aerial view of Manhattan, including parts of Brooklyn, Staten Island, Queens, the Bronx -the Greater New York that was consolidated on January 1, 1898– and New Jersey. Enormous pyramidal-shaped skyscrapers crowned with domes fill the island, two sets of twin skyscrapers linked by aerial bridges particularly stand out. The buildings are very large and seem to take up an entire block, so that there must be streets either running under or blocked by them. The width of the buildings is much greater than can be allowed by actual street blocks, and none of the streets laid out by the 1811 plan are visible.

In the first days of the following year, a drawing published in *Collier's Weekly* shows a fragment of the aerial space of New York's main avenue, as it would be

34 Image published in Dietrich Neumann (ed.), *Film Architecture. Set Designs from Metropolis to Blade Runner* (Munich, London, New York: Prestel, 1999), p. 35.

35 'The city of the future. A daring solution to the traffic congestion problem', *CyC*, September 10, 1913, no. 774.

36 Published in *Supplement to The World* (New York: December 30 1900). Copyright 1900 by The Press Pub Co. NY, cited and reproduced in Judy Crichton, *America 1900. The Turning Point* (New York: Henry Holt and Company, 1998), pp. 260–261.

in 2001: 'Looking Forward – A.D. 2001: Broadway, New York, as it may appear a hundred years hence, when modern inventions have been carried to the highest point of development'.[37] Two elements that particularly stand out in this drawing are verticality and height. Buildings are connected by air through the use of trolley-like vehicles running on high-tension cables and flying machines for one or two passengers suspended from some sort of small blimp.

Very high up, among the clouds, can be seen some traditional, but very large, electrical lights; there are also a great number of advertisements announcing great advances in communications, such as 'Wireless Telephone, Local and European,' and 'To Europe, 6 hours by Submarine' and other types of signs that allude to technological progress in relation to human life.

V. Concluding Remarks

When we compare the cities of today with the images and ideas presented in this chapter, it is evident that many of these concepts are reflected in parts of the contemporary urban landscape. It is difficult to follow the road of the successive mediations that took place between the formulation of an idea and the effective transformation of the city. Yet the influence of some of these ideas is still evident in Buenos Aires today.

An interchangeable future imagined at the beginning of the twentieth century, which was understood as appropriate for any metropolis – New York, Buenos Aires or others – could be found today in some sectors of the world's great cities. In fact this is evidence of the globalisation of ideas, ideas about the metropolitan future.

Developing a vision of the future should be rooted in understanding how people have historically thought about the future, how they expressed themselves, how the ideas entered the public domain and debate, and ultimately how the ideas were embodied in action. The significance of the research on which this chapter is based, therefore lies both in its academic value as well as in its potential social and institutional impact.

Postscript: images that refuse to die

Even if 'the vertical city' does not constitute a mimesis of the urban architecture of the twentieth century, we have found two instances in which these images were used as logos, in two occasions at the end of the twentieth century: at an urban conference in St. Louis, Missouri, in 1996–1997, and at an exhibition in Los Angeles in 2000. It would seem that they have acquired the status of the urban paradigm of the twentieth century, and continue to have some type of attraction or, at the very least, a power of representation for the century as a whole.

---

37 Published in *Collier's Weekly*, January 1901, reproduced in Critchon, op.cit. p. 2.

**Figure 5.12
'A forecast. Looking
up Olive Street. St.
Luis, Missouri 2010.'**
Source: From *Greater St.
Luis Magazine*, December,
1910. Artist unknown,
Missouri Historical Society
Archives. Published in a
brochure for the Saint Louis
Urban Forum 1996-1997:
'The Future of the City. The
City of the Future'.

Figure 5.13   Fritz Lang, 'Metropolis', Film still, 1926.
Source: Published in an ad for the exhibition 'At the End of the Century: 100 years of Architecture', held in 2000, at The Museum of Contemporary Art (MOCA), Los Angeles, CA. LA Weekly, May 5–11, 2000, Los Angeles, CA.

Chapter 6

Words and History:
Controversies on Urban Heritage in Italy

Giorgio Piccinato

Giorgio Piccinato is professor of urban and regional planning, head of the Dept. of Urban Studies, Roma Tre University (Italy). Past president of the Association of European Schools of Planning, consultant to the United Nations and the European Union. Among his publications: *La costruzione dell'urbanistica. Germania 1870– 1914* (The construction of planning. Germany 1870–1914; Rome 1974, Wiesbaden 1983 and Barcelona 1993) Un mondo di città (A world of cities), Turin, 2002.

Recent Origins

Reviewing the debate on historical city centres, as expressed in Italian literature, presupposes from the outset the identification of different currents that have fostered it. These currents are firmly linked to a period during which diverse economical and societal struggles occurred, in such a manner that the cultural debate cannot ignore them. Moreover, even if our focus is on what was achieved (and spoken about) in the second half of the twentieth century, we cannot avoid briefly examining the extent to which this period flows from previous ones. The subject before us is limited to the period following Italy's unification (1861), omitting the era of the *Grand Tour* and the first State decrees preceding the unification of Italy. These decrees already recognised the public value of artistic heritage in the 1700s and 1800s. The most prolific periods for reflection on the topic of historical centres are the ones characterised by greater economic and social activity. This is not a coincidence. It is during those years that there is the most intense pressure to transform central areas, a result of accelerated urban growth and scarce public development control.

The first period of urbanisation is that linked to the development of industrial society. In the decades straddling the late nineteenth and early twentieth centuries, a period that concludes with WWI, Italy initiated a series of changes that would alter the structure of the economy as well as the urban configuration of the country. It is the period of industrial growth, the evolution of a modern bourgeoisie and significant proletariat groups. The transformation often demolishes or guts the urban fabric, walls that still encircle the city are destroyed, boulevards and piazzas are built and new railway stations are connected to the city centre. Hygiene and traffic are the issues through which modernisation intrudes everywhere, first in capital cities such as Florence and Rome and later in the large industrial centres like Milan, Turin and Genoa. These interventions, in city areas that today we consider historical, often deal with the presence of monuments and monumental complexes whose destruction seems inappropriate.

Thus a division is born between the proponents of restoration and those fascinated by the theories of Ruskin.[1] The former, in the footprints of Viollet Le Duc with the support of eclecticism dominant in the ongoing contemporary architectural scene, gives rise to an often-fanciful reconstruction boom, principally of medieval complexes. The latter upholds the need to conserve monuments as they are and deny the legitimacy of add-ons or not strictly necessary finishing touches. An example of restoration initiatives is the decree of the Minister Fiorelli in 1882, supporting the opportunity to recuperate the 'normal' state of the monument. On the other hand an example of 'conservationism' is the Italian Charter of restoration, passed in 1883 by Camillo Boito, engineer and man of letters, on the occasion of the 4th Congress of Engineers and Architects of Rome. It is to Boito that the most frequently adopted operational mode in Italy must be attributed. It foresees the necessity of indispensable modern interventions that respect, as much as possible, the superimposition of multiple periods, thereby opting for conservation rather than restoration.[2] In this early period the professional debate is mostly around monuments, while actual practices involve the urban fabric as a whole.

It is not until 1913, that a new line of debate appears when Gustavo Giovannoni first publishes his essay '*Città vecchie ed edilizia nuova*' (Old cities and new buildings), underlining the 'diversity' of the historical city centre that for structural and social reasons cannot effectively conform to the modern city. The old city, or at least its most ancient parts, is destined to remain detached from the rhythms of contemporary life, rhythms that are better organised in newly developed areas of the city. City inhabitants too are diverse: some are young, competitive, entirely immersed in the values and ideas of modern society which unite them, while others, those of the historic city centres, are more unruffled and dedicated to old trades that have become marginal. In these last instances interventions are made solely for hygienic purposes. They are faithful to Giovannoni's theory, which envisions the restitution of the antique city fabric to its original state by excising the additions that were superimposed over the centuries, making the city fabric more spacious and uncluttered.[3] We are no longer simply dealing with the 'insertion' or 'adjustment' of new buildings to safeguard ancient ones, but with the protection of the old city centre from the combined aggressions of traffic and speculation. The leap in quality proposed by such a position is evident: the focal point moves from monuments to urban fabric, from traffic control to planning. Giovannoni's ideas became an important element in planning practice, taking into account the growing professional

---

1   G. Miarelli Mariani, *Centri storici. Note sul tema* (Rome: Bonsignori Editore, 1993).

2   C. Boito, *Questioni pratiche di Belle Arti* (Milan: Hoepli, 1893).

3   G. Giovannoni, 'Vecchie città ed edilizia nuova', *Nuova Antologia*, 995 (1913): 449–460.

and academic weight of the author himself. At the 1929 International Congress on Housing and Planning in Rome, urban planners like Stübben, Poëte, Chiodi and Piacentini concurred on this stance.[4] The new general urban planning law of 1942 confirms the two levels of planning suggested by Giovannoni in a number of his writings: one general municipal level and the other more detailed concerned with specific urban sectors. However, given the strong political influence at the time with regard to the use of historical heritage, the twenty-year period (1922–1943) of fascist planning had a largely negative result. The consequences are particularly evident in Rome, where the emphasis on a supposed continuity between the imperial city and the new regime, will be a major reason for demolishing entire blocks of buildings of medieval and baroque origins, structures not always of 'lesser' value.

Modern Architecture an Ally of Building Speculation?

After WWII the debate on historical city centres returns with a renewed intensity. In a dynamic economic context, the social disequilibria between North and South, between city and countryside, between lowlands and highlands give birth to considerable transformations in population distribution. In this period a substantive industrialisation of Italy's economy occurs. The country's urban configuration now faces pressures that had already perturbed historical city centres in other more economically advanced countries. In a context in which almost all Italian cities had suffered war damage and bombardments, reconstruction plans were hastily adopted. They often became tools to satisfy the demands of speculation more than the needs of a proper urban policy. Urban centres are beset by a seemingly runaway demand for new buildings: housing, offices (administrative and commercial). In these circumstances, inadequate public transport benefits the ascendancy of the automobile, consistent with national development policies that privilege industrial machinery and foremost that of the automobile industry. The pressure to demolish and reconstruct the existing urban fabric has an increasing number of allies among economic and political forces, because the value of areas increase, especially in city centres, with high density construction.[5]

The polemic surrounding the opportunity to erect modern works of architecture in old cities merges with the ideas of those who oppose the destruction of old city centres solely for speculative purposes. A divide arises between architects and planners: the former focused on the new city form, the latter on the manner of safeguarding the existing city. In the years succeeding the war, even with respect to fascist demolition practices and rhetoric, it seems that the division is radical: modern architecture without compromise versus integral conservation of historical heritage. Architects and historians seem to agree on these principals. However,

---

4 F. Ventura, 'Attualità e problemi dell'urbanistica giovannoniana', in G. Giovannoni (ed.), *Vecchie città ed edilizia nuova*, Florence: CittàStudiEdizioni, 1995.

5 I. Insolera, Vicende economiche-sociali e conservazione dei centri abitati, *Ulisse*, V, (1958): pp. 1454–1460.

Figure 6.1   Banfi, Belgioioso, Peressuti and Rogers: the Torre Velasca in Milan, inaugurated in 1958

architects proclaim the legitimacy of contemporary architecture within the historical urban fabric while planners contest it with determination.[6] There is no lack of incoherencies: for example the reconstruction of *Por Santa Maria*, a commercial street which leads to the *Ponte Vecchio* of Florence, which was destroyed by German forces during their retreat. In this particular case, a solution was found in a mediocre compromise: a profitable increase in density accompanied by an ambiguously amorphous architecture, a mix somewhere in between modernity and local folklore.[7] The case in point is all the more infamous in that it occurred in an urban context of international renown, which provoked harsh criticism. The awkward problem is resolved, apparently, in a decisive victory for the proponents of modern architecture in alignment with international parameters. Nevertheless, a few architects among the most noteworthy on the professional and publicist front, with curricula marked by absolute modernity, insist on finding an original solution: one that does not renounce the principals of rationalism and is not indifferent to the urban context. Architects such as G. Michelucci, I. Gardella and above all the B.B.P.R.,[8] propose an intriguing hypothesis of relationship with the so-called pre-existing surroundings.[9] The *Velasca* tower of the B.B.P.R. is a fascinating 'conservationist' response to the superb modernity of the *Pirelli* skyscraper by Giò Ponti. It seems to be a sort of manifesto.[10] Henceforth, a division seems to appear, that between architects and planners, even if they come from the same architectural and engineering schools. The distinction between those who deal with urban form and those more interested with the organisation of urban space becomes increasingly evident.

Is Conservation Leftist?

Towards the end of the 1950s, the subject of historical city centres became a dominant theme discussed in journals like *Metron*, *Edilizia Popolare* (Social Housing) and Urbanistica (Town Planning) as well as conferences of the National Planning Institute (*Istituto nazionale di urbanistica*) and Social Housing Institute (*Istituto case popolare*). More importantly, historical city centres became a popular subject, debated with great passion in municipal councils and featured in opinion magazines. In Rome and other cities, demolitions are halted as press campaigns of A. Cederna and the newborn association *Italia Nostra* awaken public opinion. In Master plans,

---

6   C. Brandi, 'Processo all'architettura moderna', *L'architettura, cronache e storia*, I/11, (1956): 356–360; and R. Pane, *Città antiche ed edilizia nuova* (Turin: Einaudi, 1956).

7   E. Bonfanti, 'Architettura per i Centri storici', *Edilizia Popolare*, 110 (1973): pp. 35–56; and U. De Martino, Cento anni di dibattiti sul problema dei centri storici, *Rassegna dell'Istituto di Architettura e Urbanistica*, II/4, (1966): pp. 75–116.

8   Banfi, Belgiojoso, Peressutti, Rogers architecture studio (Milan).

9   M. Tafuri, *Storia dell'architettura italiana 1944–1985* (Turin: Einaudi, 1986, I ed. 1982, pp. 64–74.

10   E.N. Rogers, *Esperienze di architettura* (Turin: Einaudi, 1958).

the safeguarding of historical city centres assumes an ascendant priority in debate.[11] There are two major consequences: 1) historical city centres become a regular part of the general urban development agenda; and 2) a concomitant increase in the value of heritage buildings. From this moment on market forces, which had usually lobbied for demolition and reconstruction ventures, begin to endorse conservation policies. In the post-war period town planners had sanctioned the exceptional character of historical city centres. The validation often resulted in the removal of historical city centres from general land use plans (e.g. Rome's 1962 Master plan) for ulterior elaboration in separate more detailed plans. An 'organic' law was solicited for historical centres, and potential public intervention was considered for such areas. These considerations were wide-ranging, involving themes such as the legal aspects of expropriation for public use, the creation of temporary housing for local inhabitants during urban regeneration operations and the magnitude and consequences of public sector interventions. The Gubbio Charter, signed in 1960 by renowned planners and by representatives of numerous municipalities on the occasion of the founding of the National Association of Historical-Artistic City Centres (ANCSA), is the result of wide ranging deliberations on the subject. At that stage, land speculation, the inadequacy of control measures and the natural obsolescence of old buildings were denoted as major obstacles to the preservation of old city centres. The need to identify protected areas within land use plans is brought forth, and criteria to draft detailed plans and methods for public financing and the redistribution of the resulting economic value of renovated buildings are all proposed. Finally, all restoration activities and stylistic additions are strongly discouraged.[12]

The widely publicised Master plans of Assisi and Urbino are promoted as models of intervention in historical centres. The two plans are very distinct and reflect the different schooling of G. Astengo and G. De Carlo, who are both professors at the prestigious Venetian School of Architecture. The Assisi Master plan is equipped with precise tools to analyse the physical, economic and social structure of the city and its environs. There is a clear intent to create a plan able to specify urban policies closely linked to more general social and economic development plans.[13] On the other hand, the Urbino plan is more oriented towards recognizing the value of city spaces and to building consensus, whether it is in terms of conservation of the city centre or in terms of the development of the modern city.[14] In 1970, the difference between the two approaches will be confirmed when De Carlo refuses to participate in the creation of a new planning school promoted by his colleague G. Astengo in Venice.

---

11 A. Cederna, *I vandali in casa* (Turin, Einaudi, 1957); and A. Cederna, *Mirabilia Urbis* (Turin, Einaudi, 1965).

12 Ancsa, 'Salvaguardia e risanamento dei centri storici. Il convegno di Gubbio', *Urbanistica*, 32 (1960): pp. 66–119.

13 G. Astengo, 'Assisi: salvaguardia e rinascita', *Urbanistica*, pp. 24–25 (1958): pp. 2–8.

14 G. De Carlo, *Urbino. La storia di una città e il piano della sua evoluzione* (Padova,: Marsilio, 1966).

Living in the Historical City Centre (in Bologna and Elsewhere)

The case that will make the Italian school famous will be that of Bologna. In that city, P.L. Cervellati, a young architect who had become chairman of the planning committee, applied an interventionist strategy that summarised many of the issues thus far contained in conferences and academic journals. Bologna, a communist island in a politically moderate Italy, had already been noted for some time as an example of good administration. Each significant initiative originating from Bologna had a national impact and sometimes even international, thanks to the excellent communication capability of its administrators. The fundamental idea behind the intervention was that of renewing the historical centre as a nucleus or model of a modern genuinely liveable city, somewhat closer to the expectations of city dwellers than the speculation implemented in peripheral areas. The guiding principles of their policy were: 1) active participation of city dwellers; 2) maintaining the stability of residents and of local activities; and 3) the involvement of owners who guarantee controlled rent in exchange for financial support from the municipality. Actually, this policy was applied to heritage buildings that were already public property. What makes Bologna a noteworthy example is the survey and intervention methodologies used on the urban fabric. A typological analysis is completed with the aim of identifying the components and types of common clusters. The information is then reassembled to adapt these clusters for housing purposes.[15] The same operation was carried out for load-bearing structures and façades. In fact this operation was seen as a kind of restoration or a stylistic addition, in stark contrast with the Gubbio Charter and attracted bitter criticism.

In the 1970s it is again ANCSA that revives the debate. In 1973 at the congress in Bergamo, historical centres are presented mainly as an economic asset. This presents a decisive change in perspective with several significant consequences. If cultural value is not the only element to be considered, the integrity of interventions to adjust the historic centre to the necessities of modern life and the concern for the fate of the inhabitants is also altered. This means withstanding the isolation of the historical centre tenets in favour of their full integration in municipal planning, as well as their insertion in housing policies, which also implies re-examining functions and roles of various city zones and reorganizing the territory's services and infrastructure. Subsequently, countrywide analyses demonstrated the remarkable variations existing between historical centres. Some were subject to intense development pressures and others were simply being abandoned.[16] As expected, this was linked to ongoing asymmetrical national development, defined by large internal migratory flows from the South to the North and from the country's interior to coastal areas. The idea that historical centres could play an important role in public housing

---

15 P.L. Cervellati and R. Scannavini (eds), *Bologna: politica e metodologia del restauro* (Bologna: Il Mulino, 1973).

16 Ancsa (1973), *Una nuova politica per i centri storici*, Atti del Convegno di Bergamo 1971 (Genoa: Edilart, 1973).

policies, when in the 1970s the question of housing had taken on a significant social and political preponderance, seems to draw the attention of planners. As a result, a series of housing laws, pilot projects and redevelopment plans emerge. Above all planners concentrate their efforts on defining legislative tools at the regional and urban scale.[17] The re-orientation defines an important change in social housing policy, relative to the norms generally applied in peripheral areas, where the price of available land was more affordable. The benefits obtained are twofold. On the one hand, the city centre's built environment remains in good condition and, on the other, the heterogeneity of social classes in the historical centre survives. At that time, 're-use' and 'existing city' became very popular terms amongst planners. Unfortunately, in reality the results fell short of those expected: public financing for construction was always inferior to what was anticipated and financing intended specifically for historical centres constituted but a minimal part of the proposed financing.

On Type and Form

The thematic orientation of certain architects is very different from planners. At the end of the 1950s, a new area of interest surfaces as a result of S. Muratori's and G. Caniggia's research, in which it is thought that building typology and urban morphology – not form – are the fundamental structures defining the city. Morphology is seen as the result of a collective culture matured over time, a permanent quality independent from style, which is variable because of its very nature. Similarly, building typology, rather than changing due to historical events, is thought to be closely correlated to the proper use of materials and traditional techniques. These hypotheses were substantiated by analyses of historical city centre structures, such as those in Venice, Rome and Como – studies in search of rules for 'organic' city growth.[18] The operational value of these ideas resides in their independence from aesthetics, where often too much space is reserved to diverging opinions and to cultural modes that change with time. These theses appear to offer support for restoration and reconstruction policies, such as the ones adopted with a less rigorous typological analysis in Bologna.

Typology and morphology are even the privileged field of interest of a group of architects related to the Venice school of architecture, presided over by C. Aymonino and closely associated to A. Rossi.[19] The objective of their work is threefold: 1) to reconstruct, using morphology and typology, the history and actual role of public and private places that have already determined the structure of the city; 2) to identify the architectonic 'continuity' and functionality of the city; and 3) to classify the antique

---

17 C. Carozzi and R. Rozzi, *Centri storici questione aperta* (Bari: De Donato, 1972); and Bonfanti, 'Architettura per i Centri storici'.

18 S. Muratori, 'Studi per una operante storia urbana di Venezia', *Palladio*, pp. 3–4, (1959): 97–209. G.F. Caniggia, *Strutture dello spazio antropico* (Florence: Uniedit, 1976).

19 Aymonino C. et al., *Le città capitali del XIX secolo: Parigi e Vienna* (Rome: Officina Editore, 1975).

Figure 6.2 **G. Caniggia: exercise for the architectural design class of S. Muratori (1964)**

city in formally and functionally homogeneous parts. It has a different objective than the work of Cervellati or Muratori. In short, it is the project of contemporary architecture that is the real objective of research. Architecture as intended by A.Rossi (and by L. Grassi) although defined as 'historicism' is not necessarily related to the debate on historical centres. It is developed with its own stylistic features even if it is linked to typological lines identified on a case-by-case basis.[20] Through this group's work, typology will be defined as a planning instrument for the modern city, in a certain measure independent from architectural history and styles.

Not very distant from Rossi and Grassi, both in terms of methodology and marketing, are two other different initiatives: the *Strada Novissima* at the 1980 Venice Art Biennial and the exhibition '*Roma interrotta*'. The first initiative was a project organised by Paolo Portoghesi. Several architects, critical of the International Style, were called upon to create a façade along a street within a predetermined dimensional scheme, but defined by their own individual canons. Above and beyond stylistic preferences and eclectic disparities, the determining element of this initiative consisted precisely in reasserting the 'street-alley' of the pre-modern tradition (that from then on becomes post-modern). In the second initiative, defined by G.C. Argan as 'a series of acrobatic exercises for the Imagination on the parallel bars of Memory', architects destined to become famous, put forward a planning proposal set in G.B. Nolli's eighteenth-century plan of Rome. It is a somewhat formal exercise, but nevertheless meaningful for the mode in which morphology is put forward as an element of continuity between the ancient and the new.[21] The fact remains, that on a professional level and beyond a certain elitist attitude that characterises these proposals, typological and morphological analyses (even if sometimes in less rigorous formats) have become by now common tools used in all historical centre intervention projects.

There is no lack of research that tries to extend and translate into plans the results of reflections that deny a substantial discontinuity of architecture and contemporary city from historical ones. Or better still, that rejects the value of a split imposed by the twentieth century avant-garde, underlining the rigidity that city dwellers manifest in various modes whenever there's a question of intervening in a sensitive or acceptable manner in the traditional organisation of urban space. M. Romano contends that 'because the city in Europe merges ... the very root of our identity and our human hope of immortality, the programme of not recognizing the past ... would always deprive it of sense'.[22] The author draws planning guidelines from these considerations that, on the basis of careful revision of the city structure, explicitly

---

20 A. Rossi, *L'architettura delle città* (Padova: Marsilio, 1966); and G. Grassi, *La costruzione logica dell'architettura* (Padova: Marsilio, 1967).

21 *Roma interrotta* (1978), Incontri interrnazionali d'arte e Officina Edizioni, Roma. Testi e progetti di G.C. Argan, Ch. Norberg-Schulz, P. Sartogo, C. Dardi, A. Grumbach, J. Sterling, P. Portoghesi, R. Giurgola, R. Venturi e J. Rauch, C. Rowe, M. Graves, R. Krier, A. Rossi, L. Krier.

22 M. Romano, *Costruire le città* (Milan: Skira, 2004), p. 259.

FARM HOUSE

ARC OF TRIUMPH

EXEDRA

FOUNTAIN

PROMENADE

MONUMENTAL STREET

BOULEVARD

TOWN HALL

CHURCH & BELL-TOWER

MAIN SQUARE

SKYSCRAPER

MARKET SQUARE

WALLS

SPORTING CENTRE

GATE

GATE

WALLS

WALLS

MONUMENT

PUBLIC GARDENS

MAIN STREET

PUBLIC GARDENS

Figure 6.3   M. Romano: project for a new town in Piedmont (2005)

lead back to the shapes of the nineteenth-century city. On the subject of restoration or completion in historical city centres, P. Marconi proclaims a necessary continuity between the work of the restorer and that of the architect, in the framework of a classic collective language matrix, based on the guideline of Caniggia or of Aldo Rossi.[23]

More Time, More Space

Towards the end of 1980s the topic of safeguarding historical heritage, at least in physical terms, has generally been accepted. In other words, it has been acknowledged by institutions and the media, is reflected in administrative policies and now forms part of a new social awareness perhaps even excessively so.[24] Think of the generalised use that is made of history and architectural styles, for example, in tourist villages. The re-awakening is sufficiently widespread to have fostered opposition to initiatives that can damage urban spaces of historical significance. Moreover, on a more learned level, the concept of 'historical', in the sense of identifying and protecting heritage value, applies to periods and themes unthinkable only a few decades ago. Once again, the very transformations, which cities undergo, propose new venues of rehabilitation and protection. The expansion that accompanied urban demographic growth until

---

23 P. Marconi, *Materia e significato* (Rome-Bari: Laterza, 1999), pp. 174–175.
24 B. Gabrielli, *Il recupero della città esistente. Saggi 1968–1992* (Milan: Etas, 1993).

the 1980s has slowed down. An ever greater part of construction now concerns restructuring and restoration of existing buildings, whether they be residences or offices. Older cities seem to offer a more comfortable atmosphere, certainly a more familiar one than new city areas, often devoid of 'central' places. Brick buildings, multiple residential types, streets alongside buildings and public squares, once reconverted and equipped with modern facilities, seem much more attractive today than they were 50 years ago. Factories, barracks and harbour warehouses are reconverted into offices, studios, stores and laboratories. The dismissal of industrial areas now woven into the inner city fabric, as well as the reappraisal by the market of the inner city, have expanded the limits of 'history': the nineteenth century and first decades of the twentieth century are now commonly considered to be subject to protection policies in the built environment. Similarly, there is a growing market for 'modern collections', such as furniture and other objects. Indeed, in the latest Master plan of Rome, some 1960s public housing neighbourhoods, of particular architectonic value, are now part of the protected 'historical city' category, thereby enlarging the concept of historical centres.[25] Hence, the proposal of L. Benevolo to delimit historical centres to what was built before 1870, while daring at the time, is now outdated.[26]

Not only has the period of what we call 'historical' been broadened, but also the space it encompasses has expanded to dimensions never before taken into consideration. We no longer speak of 'historical city centre', but of 'historical territory'. This augmented concept has been present for a time in Italian academic literature, starting with E. Sereni's fascinating work on landscape historicity.[27] It is certain that in Europe, and to a greater extent in Italy, there is no territory that does not have more or less obvious signs of a past history. Instead, what is new is the attempt to introduce the territorial concept into land use planning practices. The topic of landscape anthropisation, the history of agricultural land use, the relation between methods of cultivating and the structure of settlements, all become the privileged matter of study of geographers, such as E. Turri and L. Gambi, preceded by E. Bloch.[28] In this historical context, the focus is no longer solely on urban settlements: it extends to the series of transactions that have built the territory in which city centres are only an element, even if they are a dominant factor. The same applies to urban industrial features, which can also be found scattered in the countryside, in guise of hydraulic or mining activities going back to the first manufacturing era. Symbols or indications of history (and memories) are for instance the subdivisions of fields, sometimes dating back to the Roman age, cultivation techniques or traditional ways of dealing with land, including land reclamation. Similarly, forests, pastures, beaches

---

25 C. Gasparrini, 'Strategie, regole e progetti per la città storica', *Urbanistica*, 116, (2001): pp. 93–107.

26 L. Benevolo, *Roma da ieri a domani* (Bari: Laterza, 1971).

27 E. Sereni, *Storia del paesaggio agrario italiano* (Bari: Laterza, 1961).

28 L. Gambi, 'I valori storici dei quadri ambientali', *Storia d'Italia, Vol. I* (Turin: Einaudi, 1972). E. Turri *Antropologia del paesaggio* (Turin: Edizioni di Comunità, 1974).

and waterways, everything we used to consider 'nature' has become a historical mark that is now accepted as part of the collective memory of humankind. Already, in the 1958 Plan of Assisi, G. Astengo described the surrounding agricultural territory, following a scrupulous classification methodology aimed at identifying standards in land use plans oriented towards economic development and closely linked with local resources. B. Secchi, at the end of the 1980s, elaborated a Master plan for Siena, inspired by a unified approach that investigates rural areas along with urban areas, which tries to identify the fundamental 'structure' of these areas.[29] Several territorial standards tie conservation or restoration to these elements, according to an all-Italian tradition that legislates what more realistically should be determined through social consensus. Once again Ancsa provides the opportunity for an operational clarification on the use of principles that by now form part of scientific literature. In 1997 Ancsa organises a convention, where administrators, experts and the public participate, on the subject of 'a plan for the historical territory', that will generate a timely debate on the difficulties of recuperating and safeguarding the territory.[30] At this point the discussion on historical territory converges with the one on landscape.

Landscape and Identity

Landscape has a different focus depending on whether it deals with historical territory, natural environment or city sprawl. These different approaches on the same subject come from various disciplinary backgrounds that convene through this new semantic field. The most recent metropolitan urbanisation phenomena are characterised by low-density housing and an increasingly pervasive presence of settlement structures of every type (e.g. commercial, administrative and recreational). These phenomena have brought to the forefront the historicity of territory, which was previously considered residual in relation to the city. Indeed, the need to identify a method of interpreting the process of social appropriation of the territory is emphasised by the combination of elements such as: 1) the sprawl of the city beyond the boundaries that traditionally defined its history; 2) the new attention paid to areas spared by urbanisation; and 3) the expansion of the residential sphere to an ever larger scale. That mode of interpretation seems to be increasingly correlated to history in which landscape is taken as a witness. Landscape itself becomes an important factor of the contemporary city, to the point of becoming the more convincing element of the city's identity. In this interweaving of transformations, caused by economic dynamics and from the implicit or explicit cultural choices expressed by society, planners try to defend or even recover identities that seem to fade a little more day by day. Likewise, the literature on globalisation, highlighting standardisation processes that cities and regions experience under the pressure of overly powerful economic factors, has favoured the dissemination of studies that emphasise the contrasts and peculiarities of different places. At first the literature's stance was evidently

---

29 B. Secchi, 'Uno schema di piano per Siena', *Urbanistica*, 99, (1989): pp. 82–88.
30 ANCSA, *Atti del XII Congresso* (Modeno: 1997).

of a defensive nature. It has since been transformed into a more positive outlook, emphasizing the possibility to use diversity as a resource rather than an element of inadequacy. As part of this reasoning there is a debate about the identity of places, neighbourhoods, landscapes, material and immaterial cultures. History, or at least what remains of it (or what can be revived) in the memory constitutes a key tool in that it offers an almost limitless source of material thanks to the expansion of the time frame and the space to which it is applied. Therefore, historic manifestations that have sometimes fallen into disuse are newly recognised, traditions connected to old ways of life return and become fashionable again, perhaps a little artificially: festivities, dances, foods, and artisan productions.

Much is being said on 'place identity', which proposes ever more refined techniques to search out hidden, intricate and forgotten identities: places marked by history, literature, cinema or street layouts that retrace ancient pathways, or even urbanisation contained by canalisation furrows long since disappeared. As the standardisation of globalisation becomes more widespread, there is a counter reaction that attempts to adhere to contexts, difficult to distinguish, but deemed essential to truly explain contemporary phenomena. Thus cities and regions declare their diversity based primarily on history, to avoid the banality of contemporary building and infrastructure development, which makes them ever more indistinguishable. In a similar manner to what the market does when it presents substantially identical products, directed at different consumer groups. It presents the same product in a specialised manner. Today, history, once an element considered a hindrance to progress, is renowned as a facet of true modernity. To what extent this is the result of a shared social need or of a purely commercial strategy, should be carefully considered, especially bearing in mind the tourist industry's power in this field.

Tourists Demand and Historical Centres Comply

This is a very important theme for all of Europe and even more so in Italy. It is so called cultural tourism that is at the base of significant distortions in conservation practices. An important part of tourism revenue is now based on tours of minor historical centres, art cities, museums, and monuments. This revenue affects the role of urban heritage, because historical centres are seen as an excellent economic resource, superior to others and in some cases even unique. Historical centres attract tourists, tourists bring revenue, and that revenue maintains historical centres,[31] a very basic and apparently irreproachable self-fulfilling sequence. Unfortunately, this is not always the case. These fragile environments are now congested with tourists who damage heritage areas. The revenues go to tourist organisations more than to the cities themselves. In this context conservation translates into a sort of mystification process. Historical centres become part of thematic parks, where history is just a means to increase business. The tourists who crowd the art cities

---

31 International Cultural Centre, *Managing Tourism in Historic Cities* (Krakow: CityProf, 1992).

change the aesthetic experience of the city itself, undermining the very possibility of enjoying its features. This applies to tourists and residents alike: when overcrowding becomes unbearable they become hostile, such as it is already occurring in Florence and Venice. What is happening is exactly what was originally meant to be avoided. The city centre with its interlaced service networks, activities and inhabitants is being transformed into one unique mono-functional zone, a hoax on the visitor who searches for what cannot be found, after a half century of zoning in an inflexible contemporary city. Historic centres now have areas accessible only to pedestrians, housing transformed into offices or luxury condominiums, ground floors occupied by businesses and squares invaded by all types and genres of restaurants. At this point it is easy to mistake as 'historical' a shopping mall built in the metropolitan area and equipped with convenient parking areas.[32] It does not always end up this way, nor does this occur everywhere, but tourism does often transform itself from a resource to a menace. On the other hand, in certain smaller sized cities that are part of more advanced metropolitan networks, economic success has been accompanied by an unusual capacity to maintain high quality at both environmental and social levels.[33] In these cases, urban heritage is recognised as local heritage and is presented without particular pretence.

Conclusion

There is little doubt that urban heritage is of greater importance in Italian cities than in other places, with several concomitant consequences. The first is a type of enormous apprehension caused by each intervention in the urban fabric. For as much as it may be necessary to ensure continued use, the resulting debate on the course of action is deployed along lines that recall medieval battles more than punctual analyses. Another consequence is the politicisation of urban heritage. This theme has been used as propaganda during the fascist era, as a sign of good administration in the post-war period and as an obstacle to development according to those who espouse moderate political positions. Structurally, the most interesting turning point in this theme is the moment it came to encompass the whole urban setting, in such a way that the debate extended to the features and fate of the contemporary city. It was clear from the start that the question of urban heritage cannot be reduced solely to a purely stylistic choice, but that it involves ethical choices based on the principles and values of our society. This very fact has caused radical acceptance or refusal of certain interventions, which afterwards clashed with reality unfolding itself with continuous mutations and adaptations. The tentative effort made by certain architects to search for less simplistic methods of connecting to history must also been seen in this light.

32 G. Piccinato, 'Le città', in E. Cecchi (ed.), *Profili dell'Italia repubblicana* (Rome: Editori Riuniti, 1985).

33 G Piccinato, 'The happy city. Urban governance in advanced economic contexts', in F.J. Monclús and M.Guardia (eds), *Planning Models and the Culture of Cities. 11th International Planning History Conference* (Barcelona: CCCB, 2004).

The emergence of landscape and identity, themes that seem prevalent today, reflects the transformations that have occurred in spatial organisation and in metropolitan life styles. Concepts, such as urban sprawl, increased consumption (including that of tourism), global economic development and cultural standardisation, open new debates and impose the need to identify new defence mechanisms and ways of planning and design.

Chapter 7

Urban Destruction or Preservation? Conservation Movement and Planning in Twentieth-century Scandinavian Capitals

Laura Kolbe

Laura Kolbe is professor at the Department of History, University of Helsinki (Finland). She is the author of *Helsinki, the Daughter of the Baltic Sea*, editor of *Finnish Cultural History I–V* and co-editor of the series *History of Metropolitan Development in Helsinki – post 1945*. Dr. Kolbe's research is in Finnish and European History, urban and university history. Her latest research deals with urban governance and policy making in Helsinki during the twenty-first century. Kolbe is founder and chair of the Finnish Society for Urban Studies. She was the International Planning History Society's (IPHS) Conference Convenor in 2000 and is a member of the IPHS Council.

The strongest meanings of many capital cities are embodied in historic or modern buildings, urban landmarks and planned spaces dating from different historical periods. New York and 11th September have recently shown what strong symbols buildings can be (in this case the WTC towers). Castles and market places, cathedrals and churches, palaces and parliament buildings, bridges and canal systems, opera houses and theatres, trade centres and Olympic stadiums, towers and glass palaces, statues and monuments are essential in constructing urban images. The capital city is at once an imagined community and an interaction of planning history, present time and future dreams. The different systems for representing a city's identity form a complex set of social, economic and cultural relationships. They are surprisingly similar in many capital cities, as will be shown here in the Scandinavian context.

Every city has a list of 'famous sites' or 'historical monuments and milieus', places were tourists stroll to see the most typical local area. National history, local memory and urban identity are interwoven in these places and buildings. Historic layers are visible in many European capital cities; the 'old city' or historic part of the city help to give a strong identity to the urban community and are used as part of urban image building. Copenhagen underlines its old roots, having developed down the centuries from a little Baltic trading centre to a major Nordic metropolis. The oldest area has now become a shopping and entertainment centre. In Stockholm the 'must' is the Old Town with its market place. In Helsinki one goes to the impressive neoclassical Senate Square, built during the first decades of the nineteenth century by the Russian Emperor Alexander I. At that time, in 1809, Finland had been separated from Sweden. In Oslo the visit must start from the Aker harbour area and from the Akershus castle, the old royal residence during the Union with Denmark (until 1814).

Symbolically these historical centres are important and they have been restored carefully during twentieth century. It is now over 40 years since the start of legislative measures to protect historical buildings, monuments and areas in European countries and the United States. The intention of this chapter is to introduce some aspects of the post-Second World War planning and urban conservation movement in Scandinavian countries and capitals, those being Denmark and Copenhagen, Sweden and Stockholm, Finland and Helsinki, Norway and Oslo, and Iceland and Reykjavik. The main concern is in the post-war period. The issue is addressed through a two-part examination. Firstly, the pattern of 'Nordic capitals' is examined and a short presentation is made on the unique network existing between them. In the second part of the chapter, an attempt is made to examine some aspects of the municipal capital city planning and compare how, why and when the urban conservation ideology became rooted in Scandinavia.[1]

Although every capital city is unique, they have many historical features in common. Comparative study can display both their uniqueness and the common elements which they all share in the post-war period, often called 'the golden age' of metropolitan development in Nordic countries. During this period the high-rise housing activity expanded in all capitals. They began to extend beyond their municipal boundaries. The different local authorities had to co-operate in their planning activities. A more cosmopolitan, modernistic and technocratic approach to planning and architecture was having a greater impact on Scandinavian urban development. This occurred at the same time when the national economies were experiencing the start of a boom period. A steadily rising standard of living was regarded as almost a law of nature. The capitals counted upon substantial urban expansion and big investments were made in communication systems, such as improved undergrounds, harbours, motorways etc.[2]

---

1   The basic work in English on the earlier development of Scandinavian capital cities is Thomas Hall's monumental book *Planning Europe's Capital Cities*, (London: E and FN Spon, 1997) and the chapters dealing with 'Helsinki', 'Christiania' (name changed into Oslo since 1924), 'Copenhagen' and 'Stockholm'. On urban development in Scandinavian countries a solid all-Nordic research was conducted during the 1970s: G.A. Blom (ed.), *Urbaniseringsprosesser I Norden I–III*, (Oslo: Universitets Tryckeri, 1977) with contributions from all five countries. More of the cultural and political pattern of 'Norden', meaning the North of Europe or Scandinavia in Oysten Sorensen and Bo Stråth, Introduction: The Cultural Construction of *Norden* in O. Sorensen and B. Stråth (eds.), *The Cultural Construction of Norden* (Oslo: Scandinavian University Press, 1997). Also Leif Anker and Ingalill Snitt, *Our Nordic Hertitage. World Heritage Sites in the Nordic Countries* (Kristiansund: KOM förlag as, 1997).

2   The tradition of writing a local history is old in Scandinavian countries and that also concerns capitals cities. Stockholm, Helsinki and Oslo have municipal Historical Committees

A central issue here, when we talk about capitals, is the special relationship between the state, the municipality and public opinion. It is obvious that at every level of the urban Scandinavian history the three players have a relationship when it comes to capital planning, the 'urban question' and the image of the 'number one city' of the nation. The status of planning is high in every Scandinavian country. Public control exercised by a professional bureaucracy is well established in the North and has been so at least since the seventeenth century. During the nineteenth century an effective system of municipal administration was created. Over the last 50 years political life in Denmark, Norway, Sweden and Finland has been dominated by liberal or social democratic parties with a great interest in the social control of land use and planning. Municipal planning monopoly in Scandinavian cities means that legally binding physical plans must be approved by the municipalities before being ratified. The control exercised is essential because municipalities are also the main land owners. Still, when it concerns individual buildings, the landowner has retained considerable power despite the municipal planning monopoly. The complex power relationship between municipal planning and public concern is one central aspect in local urban history.[3]

I. Changing the City – Horizontal Relationships Between Scandinavian Capitals

International exchange has long played a big part in the history of urban planning. A great deal has been written about the 'international planning milieu' especially by the turn of the twentieth century, before the First World War, embodied in congresses, study trips, associations, exhibitions and competitions (urban planning and architecture). Many innovatory elements in modern urban planning became a part of international planning thought, such as the renewal model of Paris under Baron Haussman, the English garden city idea, the influence of Camillo Sitte and the city beautiful movement. Lot of research is done on the big events of this milieu. No so much scholarly investigation has been directed to the networks of information between various cities, municipal councillors and politicians, planners and practitioners. Works on the 'big names' of town planning and on 'great cities' have prevented us from seeing the richness of different organised and free urban networks in Europe and also on the transatlantic and colonial level.[4]

---

working in close cooperation with the academic historians at the university. They have many publication series and publications of the capital cities' history. Many of these works hold important positions among national urban history works. More on the theme Jan Eivind Myhre, 'The Nordic Countries', in Richard Rodger (ed.), *European Urban History* (London: Leicester University Press, 1993), pp. 171–190.

3 Thomas Hall (ed.), *Planning and Urban Growth in the Nordic Countries*, Studies in History, Planning and the Environment 12 (London: E and FN Spon, 1991).

4 This theme was inspired by the classical research by Professor Anthony Sutcliffe, *Towards the Planned City. German, Britain, the United States and France 1780–1914*

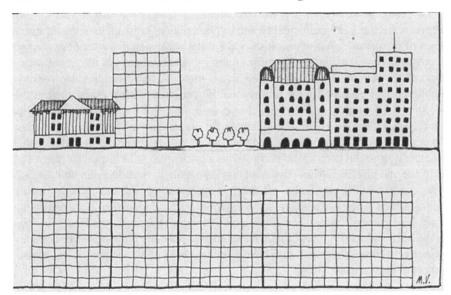

Figure 7.1   Helsinki
Source: Kolbe

When urbanisation was at its liveliest at the end of nineteenth century and before the First World War, it became clear that cities had started to form their own over-national network. The civil servants, policy-makers and architects in the Scandinavian capital cities have closely co-operated with each other since the early 1900s. Nordic conferences and congresses, meetings, close networks, study journeys, publications and personal contacts were (and still are) important and regular between urban civil servants, politicians and other civic organisations in order to answer the 'urban question'. National capitals of Scandinavian countries have build a network and developed their own foreign polices already at the start of the twentieth century. In 1923, a high level conference between the representatives of Nordic capitals took place in Stockholm. Fourteen years later, in 1937, a similar meeting was held in Copenhagen. During this period Nordic collaboration had made some progress, partly because of the growing threats from outside. The Second World War promoted a new sense of community among the Nordic countries which had not existed before, thus laying the foundations for the co-operative conference between the capitals. The next meetings took place immediately after the war, in Stockholm in 1946 and in Helsinki in 1948.[5]

(Oxford: Basil Blackwell, 1981). At the Scandinavian and North European level Marjatta Hietala, *Services and Urbanization at the Turn of the Century. The Diffusion of Innovations.* Studia Historica 23 (Helsinki: Finnish Historical Society, Gummerus Oy, 1987).

   5   Due to the tradition of writing the urban histories in the context of national history very little emphasis has been placed on the exchange of the international know-how. See Laura Kolbe, Huvudstadssamarbet i Skandinavien 1923–1946 och den nordiska myten om

Since the 1940s, these conferences of Nordic capital cities (NCCC) have been held on a three yearly basis. These conferences can be called as organised, collective entity, with their own rituals, hierarchies, rules and values. NCCC have become a dynamic institution, and an essential part of late twentieth century Nordic co-operation and feeling of unity. They are points of contact where common urban issues are discussed and compared, and where the diffusion of innovation is emphasised. The themes discussed have often had an administrative, technical, social, political or economic character, having a great influence on municipal urban planning and building in each capital. Until the 1990s, every conference was well documented and they form a valuable source for how the 'urban question' was answered during different decades. The NCCC forms an interesting bridge between the local and international; it is an antidote to national urban histories, avant-garde movements or 'great men' within planning history. To study the history of NCCC helps to understand how urban knowledge, local powers and municipal practices are transferred in between cities and metropolitan areas.

II. The Municipal Planning and Conservation Movement

The official story of post-war capital planning in the Scandinavian countries was clearly dependent on the needs of modern society. The post-war years became a period of national reconstruction and cultural reorientation of modern planning. The conservation movement was a reaction to the results of this dramatic modernistic post-war transformation. Both movements form a central aspect of late twentieth century urban development. One essential question is how modern capital planning and post-war attitudes of discontinuity relate to the conservation movement and attitudes of continuity in history? The Scandinavian capital cities provide an excellent possibility to analyse this complex relationship between planning, progress and modernism and conservation, heritage and nostalgia.[6]

In the planning history of every city, there is a moment of transition into 'the modern'. In the Nordic capital cities, that moment is closely linked to the planning

---

det folkliga självstyret, to be published during 2005 at the internet (www.helsinki.fi/norden). The meeting proceedings have been published independently by the City Governments since 1923 in Swedish, Danish or Norwegian (depending on the host town). The first is to be found in the City Archives of Stockholm *Redogörelser för förhandlingarna finnes tryckt i bihang nr 44/stadskollegiets utlåtanden och memorial, 1923* and the next in the City Archives of Copenhagen, *Beretening om den skandinaviske kommunalkonference i Kopenhavn vid midsommer 1937*.

6   In Helsinki the main research literature is based on the new history of three volumes on the capital city in the period 1945–2000 with different urban researches. Other good basic works are Steen Eiler Rasmussens, *Kobenhavn. Et bysamfunds saerpraeg og utvikling gennem tiderne* (Copenhagen: GADS Förlag, 2001), Lars Nilsson (ed.), *Stockholm – Staden på vattnet I–II* (Stockholm, 2002), Matti Klinge and Laura Kolbe, *Helsinki – The Daughter of the Baltic Sea* (Helsinki: Otava, 1999). *Oslo bys historie 1–5* by different academic authors was published during the 1980s. The two last parts deal with twentieth century development.

of a modern rail traffic system and underground/metro. Development before 1914 has many common features. An administrative tradition of civil servant rule, an economic structure still oriented towards agriculture, a lack of capital, and a slow industrial and logistic take-off caused the urbanisation process to begin late. This set-up situation changed when in the 1860s and 1870s cities started to grow and urban functions and structures became more specified.[7]

All Nordic capital cities developed gradually in the period between 1870–1914 to become the cultural and political centres of their countries, the real capitals or metropolises. All cities started expanding beyond their existing borders, when the first major industrial districts and ports came about. The reforms of the municipal administration at the end of nineteenth century marked a change in urban planning. The new city council consisting of liberal, enlightened and influential men of the community approved the urban plans and the construction rule. The city's administrative court became the central administrative body. The pressures caused by traffic, new forms of production, social groups, trade, the educational system and new urban life styles influenced planning in the late nineteenth century. Urban functions developed faster than urban or traffic planning.[8]

The idea that a town should be planned before it grew – not after it – gained ground. Influence from the European metropolises was clear. In line with European models, those coming especially form Germany, the planning of the capital city underlined technological modernity, aesthetic dimensions, urban intimacy, historical and organic continuity instead of the regularity and centrally governed patterns of earlier periods. The shift from the central to the local level expressed itself in people's attitudes. The rapid changes of city centres gave birth to historic nostalgia; City Museums and archives were established to document and photograph the 'disappearing cities'. The members of the bourgeois and intellectual elite established the first local historical associations.[9]

---

7   Lars Nilsson, 'A Capital Floating on the Waters: The Development of Stockholm in a Long-Term perspective', in Lars Nilsson (ed.), *Capital Cities. Images and Realities in the Historical Development of European Capital Cities*, Studies in Urban History 22 (Stockholm, 2000), pp. 152–170. Marjatta Bell and Marjatta Hietala, *Helsinki – The Innovative City. Historical Perspectives* (Helsinki, 2002), pp. 77–99.

8   A general comparative presentation in English is in Thomas Hall, 'Is there a Nordic Planning Tradition? in Thomas Hall (ed.), *Planning and Urban Growth in the Nordic Countries* (London: E and FN Spon, 1991), pp. 247–251. Klinge and Kolbe, *Helsinki – The Daughter of the Baltic Sea*, pp. 43–72. Rasmussen, *Kobenhavn*, pp. 217–244. Knut Kjeldstadli, *Den delte byen. Oslo bys historie 4* (Oslo, 1990), pp. 337–390.

9   No comparative historical research has been done on the history of City Museums and City Archives. Arve Skivenes (ed.) *Evidence! Europe Reflected in Archives. European Cities of Culture 2000* (Santiago de Compostela: Agencia Grafica Galleca, 2000). See a local description of an urban photographer Signe Brander, who documented during the first years of the twentieth century on behalf of the City of Helsinki the disappearing urban landscape Jan Alanco and Riitta Pakarinen (eds), *Signe Brander 1869–1942. Helsingfors fotograf* (Helsinki: City Museum, Frenceliin kirjapaino Oy, 2004).

The first phase of preserving the old town environments and building complexes occurred at the turn of the twentieth century. Reformist architects emphasised the idea of preserving the historical city centres and monuments. The change in attitudes happened at the same time when the national cultural movements were strong, with mixed elements of regional nostalgia, history, folklore, vernacular and urban heritage combined with the idea of Nordic nature and local history as a source of inspiration. During the period between 1914 and 1945, the first level of national legislation and the legal basis for preservation was created. The first Royal or National Historic Monuments and Building Preservation Acts were politically approved by the national parliaments. The main problem was that the financial support from national governments was minimal.[10]

The First World War shattered the old world, destroyed cities and gave birth to new national states. The only change that occurred in Scandinavia was that Helsinki remained the natural capital when Finland separated from Russia and became an independent republic in December 1917 in the shadow of the war and the Russian revolution. Among the large problems to be solved in every capital city in the 1920s and 1930s were the matter of the railways, the housing problem and the planning of a new city centre. Part of the urban development was to annex land to a city. This process continued in Helsinki and Oslo, as well as in Copenhagen and Stockholm.[11]

With the exception of war times, the Scandinavian capital cities grew each year for over a hundred years, up until the late 1960s. For post-war Europeans in 1945, the need for physical reconstruction and political renewal was obvious. New ideas within architecture and town planning were important in producing the new forms of urban life. The idea of post-war planning legislation was to improve the efficiency of local development planning, to extend its range beyond existing densely built areas, and to create instruments for structure and general planning at the municipal and regional level. In all Scandinavian capital cities 'dynamic community development'

---

10 A comparative analysis of the Nordic heritage legal development is described in the Finnish book of building protection Aarne Heimala (ed.), *Rakennusmuistomerkkimme ja niiden suojelu* (Porvoo: WSOY, 1964). Denmark was the pioneering country in year 1918 with a state level preservation act, listing 637 buildings in Copenhagen. In Sweden the year 1920 is crucial for the Building Protection law. At the same time a list was made with 400 state owned buildings and 3,000 churches to be preserved. Norway followed in 1920, following the Danish model (with Building Acts later in 1947 and 1959) and Finland with its Building Act in 1932 and Building Protection Act in 1964.

11 Stephen V. Ward, *Planning the Twentieth-Century City. The Advanced Capitalist World.* (West Sussex: John Wiley and Sons Ltd, 2002), Scandinavian part pp. 66–67.

Figure 7.2   Oslo
Source: Kolbe

and traffic planning were combined. Nordic policies were to combine (social) democratic ideals, traffic efficiency and modern metropolitan image.[12]

The second phase of preservation movement was followed up by the general 'wake up' of attitudes during the 1960s, especially after the cultural revolution of 1968. Monumental modernistic city planning was criticised and gradually replaced by considerations of the architectural and milieu values of cities. We can see different development levels:

1. first wave of urbanisation and awakening of public opinion (1900–1910)
2. first efforts to create a heritage legislation (1920–1930)
3. post-war large scale re-development plans and listing of buildings and historic environments (1950–1960)
4. reaction to demolition of the city centres (1960s)
5. local pressure groups are formed (1960s)

---

12 A general description of the metropolisation development in Laura Kolbe, *Helsinki kasvaa suurkaupungiksi. Julkisuus, politiikka, hallinto ja kansalaiset. Helsingin historia 3* (Helsinki: Edita Oyj, 2002); and E. Cullberg et al., *Stockholm blir storstad 1945–1998* (Stockholm: Centraltryckeriet AB, 1998). See also Peter Hall, 'Social Democratic Utopia. Stockholm 1945–1980' in Peter Hall, *Cities in Civilization. Culture, Innovation and Urban Order* (London: Phoenix, 1998), pp. 842–880; and Ward, *Planning the Twentieth-Century City*, pp. 201–202.

6. preservation becomes a political issue with media support (after 1968)
7. conservation movement gets wider support (1970s)
8. general city plans, the idea of preserving the historic town and second efforts to create heritage legislation (1970s)
9. municipal management of maintenance and repair (1980–1990).

1. Nordic capital cities and post-war re-development plans

During the Second World War surprisingly little physical destruction occurred in Nordic capitals. Norway and Denmark were occupied by Germans, and Iceland by British and later American armed forces. Sweden remained neutral. Finland had its separate wars (1939–1940 and 1941–1944) against the Soviet-Union. Although aerial bombing was used on a large scale against urban centres in Europe, the capitals in Scandinavian countries were not seriously damaged. In all Scandinavian countries war fostered planning in many ways and strengthened the power of governments and municipal governance. In the four biggest Nordic capital cities, Copenhagen, Stockholm, Helsinki and Oslo, the overall planning was dominated by the idea of a greater city in terms of one big municipality. The urban regions grew in satellite communities built along the railways and roads.[13]

In the post-war period municipal planning became central in strengthening the image of the capital. The idea of post-war planning legislation in Scandinavia was to improve the efficiency of local development planning, to extend its range beyond existing 'densely built areas' and to create instruments for structure planning at the municipal and regional levels. All Scandinavian capitals established new urban and/or traffic planning bodies and made municipal plans in the 1950s based on the comprehensive analysis of both public and private needs. In these plans the systematic transportation ideology and the great urban renewal of the city met each other. With the creation of the city's public transport department, the arrangement of suburban traffic and transport became a hot political issue. These themes were already present in the first NCCC meeting in Stockholm in 1946.[14]

In all cities metro and traffic planning became highly political issues, especially within the strong social-democratic parties with great interest to the social control of land use. The traffic planning also showed an important social message: the need to be a democratic metropolis and efficient capital city. In Copenhagen there were several general plans, the Finger plan (1947) for the development of greater Copenhagen, the preliminary General Plan by the municipality of Copenhagen (1954) and a regional plan approved by politicians (1962, 1973). Stockholm is a prime example in the

---

13 Harry Schulman, 'Helsinki becomes a metropolis', in *Quarterly from the City of Helsinki Urban Facts* 2 (2000): pp. 5–9.

14 City Archives of Stockholm, *Stadskollegiets utlåtanden och memorial, Bihang nr 6/1946.* The five main themes dealt with (1) capital cities and their regions, (2) housing issues, (3), the problems within the suburban traffic, (4), youth- and children policies and (6) public financing at the municipal level. Kolbe, *Helsinki kasvaa suurkaupungiksi,* pp. 221–224.

North, and in Europe as a whole, of the vast transformation of a city centre under the auspices of the municipality with its structure plans *City62* and *City67*.[15]

Oslo and Helsinki shared many common features. Both cities were merged with the neighbouring municipalities (1946, 1948), with new urban and suburban areas for planning and housing. Municipalities in both capitals prepared a general plan during 1950s, based on zoning, suburbs and regional centres. The decentralised model – new residential areas were connected to a number of new local centres around the city centre – needed good traffic communications. An underground with rapid rail transport was planned and realised in both capitals. The Master Plan for Oslo (1960), metro plans for Helsinki (1956, 1964) and Smith and Polvinen Traffic Research (1968) reflected the general spirit of the 1950s and 1960s: a complete clearance of the old blocks and a comprehensive redevelopment, high plot ratio, motorways through central areas, clearance and redevelopment of old urban structure replaced by technically and hygienically defective modern housing. Reykjavik's Master Plan (1962–1982) was also a region-wide land-use and traffic survey. The traffic planning spirit was strong. The Master Plan declared that 'every household should have its own automobile' – and in this spirit the Plan had three highways running in an east-west direction of the city.[16]

During the 100 years, from 1870 to 1970, the Scandinavian capital cities had been remodelled from small (mainly) timber towns to continental stone cities. The years of rapid economical change between the late 1940s and the mid-1970s especially meant a heavy transformation of the urban structure and townscape. In all Nordic capital numerous old buildings were destroyed to make way for modern constructions, usually close to the central traffic areas, like railway stations and main thoroughfares. The key word was renewal (*sanering*) of many parts in the old city centre. In the city centre considerable building activity was concentrated with new offices, business premises etc. A new dimension was present in the capital planning: big-city dreams were combined with the desire to attract private business and companies. Given an attractive centre, business and other activities would be drawn to a city; without such a centre the business would go to a rival place.

2. Conservation movement – preservation as priority

The municipal policy of clearance and redevelopment, as it was being practised in Nordic capitals, gradually roused ill-will and opposition. In this regional network it is interesting to study how partly national and partly international actions to preserve

15 Ward, *Planning the Twentieth-Century City*, pp. 206–209. All articles in Cullberg et al., *Stockholm blir storstad 1945–1998* discuss the effects of traffic planning to urban development from various points of view.

16 Erik Lorange and Jan Eivid Myhre, 'Urban Planning in Norway' in Thomas Hall (ed.), *Planning and Urban Growth in the Nordic Countries* (Oxford Studies in History, Planning and the Environment 12: E and FN Spon, 1991), pp. 147–150 and Laura Kolbe, 'From a provincial to a national centre: Helsinki', in David Gordon (ed.), *Planning the 20th Century Capital City* (forthcoming 2005: Blackwell).

'heritage' influenced planning in Scandinavian capitals. This process included two aspects: the evolving framework of legislation and a changing social, economic and cultural concern on architecture and urban landscape. The question of conservation first appeared in the NCCC Reykjavik agenda in 1973 under the title 'Preservation of buildings with historic and cultural interest'. Six years later the same issue was discussed in the NCCC's conference in Helsinki 1979 and the theme was now 'Clearance and renovation of historic quarters'.[17] A clear shift occurred from separate buildings to city quarters. What had happened in the planning of Nordic capitals during these years?

Identification of the first phase – the rise of public interest in conservation – is obvious. It had already occurred during the post-war period, initially responding to the concerns of a minority elite within the field of intellectuals, urban culturally oriented middle-classes and specialists within the field (museums, heritage societies, local historical associations). Attention was focused on threats to specific buildings and monuments. Public opinion started to change as some central urban buildings from the nineteenth century were threatened. Demands for protection started to be heard, although the general attitude was still in favour of modernism in planning and architecture.

In all capital cities some municipal efforts were carried out within the field of urban history and heritage. Some city planners and researchers on the conservation of historic city structure listed and registered central historic buildings. In Oslo, in 1956, the municipal heritage committee and a civil servant made the list. In Stockholm the same occurred in 1959 when the listing for the conservation of historic buildings and manor houses identified almost 300 buildings as being of historic value. In 1965 a new list of historic buildings was published by the City Museum. In Helsinki a similar list was made by the town planning section of the city's real estate department in 1962, with some additions in 1965. The result was a list of 400 historic buildings with value for preservation. Copenhagen followed in 1966 with a list of historic buildings (600 state + 700 city defined) and Reykjavik in 1968 (conservation plan as a part of *Generalplanen* work).[18]

One central issue was economic. Even if there was enough political support for the preservation of historic buildings, there was a lack of public finance to do so. In 1951 Stockholm started to reserve public resources for the preservation of buildings with an historic value. One of the main issues was the question of the Old town or inner historic district in the capital city. Stockholm and Helsinki offer interesting cases. In the 1950s, however, certain civic and heritage organisations had already expressed their concern about the oldest urban milieus. In 1952, the City of Helsinki

---

17 City Archives of Helsinki, *Referat fra den Nordiske Hovestadskonference I Reykjavik 1973* (Reykjavik, 1973) and *Redogörelse för det Nordiska Huvudstadsmötet I Helsingfors* (Helsinki, 1981).

18 See above, the *Referat fra den Nordiske Hovestadskonference 1973* the participants give information to the meeting of the measures undertaken by the municipalities thus far, pp. 49–64. This presentation is based on that text.

Figure 7.3   Stockholm
Source: Kolbe

declared the Senate Square and its surroundings the historical centre of the city, which should be preserved. After the Independence in 1917, pressure had already increased for the redevelopment of the historic core. There were groups of people who wanted to preserve the remains of the old townscape, already altered because of the rapid urbanisation. As the examples of Europe showed, the identity of many historical cities was closely connected to the historic core.[19]

Today in the histories of these cities, the Old Town (Stadsholmen) of Stockholm and the Senate Square of Helsinki have retained their prominent role. The Old Town in Stockholm, built mainly in the 16th Century, provides living evidence of a truly long historical development as a capital city. In this part of the city, or at close proximity, the castle, main church, the parliament, the bourse and House of the Nobles are located, as well as many historic palaces. It is of central value when the urban history is told. In Helsinki the Senate Square and the Market Place, as well as the quarters around the main squares, have kept their prominent role in the city since the early years of nineteenth century (1810–1852) when this area was created. The neoclassical buildings in the heart of the newly established capital are used as the basis to describe not only the urban development, but also national history.[20]

Symbolically, the Senate Square became the centre of independent Finland after 1917. Politically the connotations of the area reminded of the centralist rule of the Russian imperial tradition. When transforming this scene into national core, the architectural historians indicated to the aesthetic connotation with Greek and Roman Antiquity, German Classicism and the Russian Empire Style, and its intellectual links with Enlightenment thinking. In this way the square and its architecture could be linked to the great European tradition. Senate Square had its own identity and it could be compared with any plans of the same period. In this sense the planned square and its architecture could be accepted as a national symbol for a republican nation. Finland could thus demonstrate that it was on the same cultural level, a nation equal to other civilised nations.[21]

Stockholm, the capital city of the Royal Kingdom of Sweden, since the fifteenth century makes its urban history so that it underlines Stockholm's 'preparation' for becoming a capital city not only for Sweden but also for Finland (until 1809) and many other regions at the Baltic Sea during the seventeenth century. Pride in the area today called Old Town (Gamla stan) and its role in the local political history plays a key role in the Swedish narratives: the main data and history of the Old Town as the symbolic core of the expanding city – and the kingdom – is presented with special emphasis on how the city developed its functions as a capital. Usually here the urban

---

19 Anthony M. Tung, *Preserving the World's Great Cities. The Destruction and Renewal of the Historic Metropolises* (New York: Clarkson Potter/Publishers, 2001); and Francoise Choay, *The Invention of the Historic Monument* (Cambridge: Cambridge University Press, 2001).

20 Göran Dahlbäck, 'Stockholm blir en stad' in Lars Nilsson (ed.) *Staden på vattnet. Del I, Stockholm 1252–1850* (Stockholm: Centraltryckeriet AB, 2002); and Klinge and Kolbe, *Helsinki – The Daughter of the Baltic Sea*, pp. 23–32.

21 Anja Kervanto Nevanlinna, Refined monuments. Transformation of the historical heart of Finland, in *Quarterly from the City of Helsinki Urban Facts* 2 (2000), pp. 12–14. See also Sinikka Vainio (ed.), *Kaupungin Leijona-sydän, Stadens leijonhjärta, The Lionheart of the City* (Helsinki: City Museum, Gummerus kirjapaino Oy, 1998) describes the history of the old houses in the heart of Helsinki, which today occupy the city's central administration.

history is described as 'Western' and thus Stockholm is placed among the leading capital cities in the northern part of the Europe.

The medieval age is a central historical period and is often underlined, if possible, in the urban histories. It is a period when the geographical entity of the modern Swedish kingdom was created. The period is synonymous with commerce, strong town culture and merchant's cultivated values and manners. German language and the Catholic Church were unifying factors in North and Central Europe. The medieval merchant represented a mentality of self-reliant independence from royal and aristocratic powers. The cities had their town halls, walls and towers and armies, and the merchants had their own churches. This history is well presented especially in the Old Town of Stockholm.[22]

The process in preserving these old parts of the city was similar in Stockholm and Helsinki. The only specific measure of building protection to take place during the 1950s was the protection of the monumental environment around the Senate Square in the 1952 Olympic year, on an initiative of a special archaeological committee. The square was declared an 'old district'. In Stockholm, a committee was formed in 1960 for analysing the possibilities to modernise and renovate the Old Town. After two proposals (1962, 1965) the committee underlined the value of this old part of the city. It recommended that only in exceptional cases old buildings could be destroyed. A recommendation was made: the area should be restored in 16 years. City municipalities gave political support for preservation in two stages, in 1967 and 1970. The state was expected to give support to the preservation.[23]

In Copenhagen the first reduction of building rights was made in 1959. Helsinki followed in 1961: the sale of permits to additional buildings was limited in the city centre. In the 1960s the attitudes changed. During the decade there were conflicts over whether or not to preserve certain valuable buildings – and soon even certain culturally valuable sites. In the second phase of conservation movement local elite concerns gradually developed into a radical movement. This is to be seen at the end of 1960s. A critique towards the results of modern city planning grew everywhere in Western Europe. This dislike of modernism was part of the global 1968 movement. Certain traffic plans or local authority plans to clear old urban buildings, parts or parks were strongly opposed by students and intellectuals – by new 'democratic forces' of the society. A series of urban anti-development books and pamphlets were published from the late 1960s onwards. The ability of this movement to voice its concerns through the media was great. Here an increase in conservation due to international influence, UNESCO's initiative for the International Year of Historic

---

22 The 'official urban history' of Stockholm was rewritten for the year 2002, for the city's 750s jubilee. See Dahlbäck, 'Stockholm blir en stad' and Robert Sandberg, 'Huvudstad i ett stormaktsvälde' in Lars Nilsson (ed.) *Staden på vattnet. Del I, Stockholm 1252–1850* (Stockholm: Centraltryckeriet AB, 2002).

23 A description of this process was given in Reykjavik in 1973, *Referat fra den Nordiske Hovestadskonference I Reykjavik 1973*, pp. 175–177.

Figures 7.4   Käpylä old area
Source: Kolbe

Figure 7.5   Käpylä new plan
Source: Kolbe

Monuments (1964) and the founding of the European Architectural Heritage Year (1975) should also be noted.[24]

'The Swinging Sixties' became soon a decade with political unrest. In the spirit of many demonstrations (against the war of Vietnam, working conditions, housing policy professional powers-that-be at the universities etc.) some groups, squatters, occupied condemned properties. Young occupants demanded influence over redevelopment, planning and building. In every city one can see, that the public opinion finally changed under similar circumstances. Capital city media – newspapers – played a central role. It all culminated in the years 1969–1971.

In Helsinki the preservation of the old municipal wooden garden city of Käpylä came to a decisive moment in these years. In Oslo the hot topic concerned the preservation of old urban parts Rodelokka, Vålerenga and Kampen. In Reykjavik 1969 a central political debate circulated around historic areas in the city centre (east side of Laerkgatan). In Stockholm the fight for alms in 1970 in the central park area of Kungsträdgården (building a metro station below) was a turning point. The Free City of Christiania, the occupation of the former Bådmandsstraede military barracks in Christianshavn in Copenhagen in 1971, came to symbolise the youth revolution.[25]

The structure of the active pressure groups was also similar. Helsinki being an old university town, the student organisations and some professors, Helsinki-Society, Museum for Finnish Architecture, local urban societies (Käpylä-Seura), pedestrians (1968: Enemmistö ry) were central. In Oslo the core of the critical group was formed by university intellectuals in the 1968 movement, journalists and architects, as well as by the local historical associations. In Stockholm the pressure groups were also formed around St. Erikssamfundet, city museum, as well as left wing artists, intellectuals and students and the 'Alm movement'. In Reykjavik the national association for architects was active (Isländska Arkitektföreningen). Also in Copenhagen the same pattern existed: Foreningen til Gamle Bygningens Bevaring (Society for Restoring the Old City), cross-political local societies, (Christiania: '24 person of high cultural position') and many intellectuals, artists and students were active. Typical for the period were many occupations, demonstrations, scandals and

---

24 Kolbe, *Helsinki kasvaa suurkaupungiksi*, pp. 273–293 describes in detail this change in Helsinki, and makes a comparison with other Scandinavian capital cities.

25 Thomas Hall, 'Stockholm planerar och bygger 1850–2002', in Lars Nilsson (ed.) *Staden på vattnet. Del I, Stockholm 1252–1850* (Stockholm: Centraltryckeriet AB, 2002), pp. 261–287. Laura Kolbe, 'Will the wooden Käpylä be saved? The beginning of a conservation movement in Helsinki', in *Quarterly from the City of Helsinki Urban Facts* 2 (2000), pp. 21–24. The history of the Christiania case has been described in many books, articles, films, and television programmes. See more on www.christiania.de on the history and reference literature. The heroic aspect is evident as this quotation from the Internet shows: 'Christiania's over 30-year history is a long and tangled tale of struggles, victories and defeats. Many of the people who were in on the start of the experiment are no longer with us. But the dream of a life in freedom, and the idea of a city ruled by its inhabitants, is still alive. People from far and near are still attracted by the Freetown's magic mixture of anarchy and love.'

Figure 7.6 Christiania
Source: Kolbe

outrage among the rest of society. In this sense the Free City of Christiania was unique. It became a symbol of protests against existing social and cultural norms.[26]

The city administration and policies took up this matter for discussion more slowly. The majority of politicians shared the technocratic view that it would not be worth repairing old parts of the cities. Therefore it was no surprise that NCC discussed the preservation issue in 1973. The capitals reported their present local situation. Little had been done although some aspects were clear:

1. *Political restriction*: The linkage between the conservation movement and municipal planning was still weak, but this relationship was now taken seriously by politicians.
2. *Legislative restriction*: The legislation was undeveloped and could not prohibit the demolition of historic buildings and quarters which were privately owned.

---

26 So far this part of the (comparative) micro history has generally been neglected in the general 'urban growth and planning histories' of the capital cities. The issues of restoration and preservations offer interesting 'key holes' to the municipal decision making processes, to the continuities and changes, to the interaction between tradition and progress, municipalities and citizens.

3. *Practical restriction*: Holistic concern with heritage issues was still weak but the concern for historic quarters was growing.
4. *Economic restrictions*: financial support from local and national government was minimal.
5. *Cultural restriction*: modernist movement was strong in all countries – 'A living city is not an artificial museum' – attitude was strong.[27]

Conclusions

Post-war preservation movements usually started when an intellectual or artistic minority started to mobilise discussion. Soon, the local residents, too, became active. At this stage, protectors of historic buildings or valuable areas represented the 'enlightened' social stratum, including academic researchers and experts on architecture not connected to the political parties. After 1968 the public opinion started to change. Some new actors, youth, students and the journalists, gave their support to the restoration and preservation plans. Many experts were featured in the press and on television. The spirit of the 1968 movement was visible in the debate on the essence of urbanism. The years 1968–1970 meant a cultural and ideological re-orientation in the 'urban question'.

The 1980s saw the renaissance in religious and national feeling, and with it a growing respect for historic buildings and antiquities. Many political decisions were made to revive the historical parts of inner cities and master plans started to list historical milieus worth preserving. The enthusiasm for restoration expressed a completely different historical approach to that of progressive modernism in the 1960s or the earlier urbanism of 1940s and 1960s. The recent decades have been characterised by large-scale restoration work in the historic districts of the Scandinavian capital cities. The 1990s urban renewal work has been done with considerations for the environment and ecology. In contrast to earlier times, more properties are being preserved and modern dwellings fitted out with up-to-date installations behind the old facades. In Oslo, Copenhagen, Stockholm and Helsinki the oldest inner city areas have now become shopping and entertainment centres that attract tourists and people form the outskirts. In 1990s there has been a thriving cultural challenge to revalue the historic past, in connection to the commitment of Copenhagen (1996), Stockholm (1998) and Helsinki (2000) as Cultural Capitals of Europe.

Much of the destruction of historic urban fabric in the 1950s and 1960s can be attributed to the ideals of modernism in planning and architecture. But this change and rapid modernisation, post-war reconstruction can also function as identity factor, just as the inner part of Stockholm show. Today, in 2005, the values of modernist heritage are discussed in the Scandinavian capital cities. The foremost examples of modernistic architecture in these cities were often designed by internationally

---

27 *Referat fra den Nordiske Hovestadskonference I Reykjavik 1973*, pp. 49–64.

esteemed architects, and are now under revaluation. Many modern buildings are recognised as masterpieces by the international community of architects. In this sense, some central modern buildings and areas have become monuments of architectural history. From the point of the preservation tradition, the question asked today is: can they have historical value? The preserving of monuments of modern architecture is already a hot political debate in Stockholm, Helsinki, Copenhagen and Oslo.[28] Here a comparative analyse and further research could indicate how historic and modern architecture is used as a part of symbolic language in urban context.

28 Papers from the international ICOMOS conference were published in *Dangerous Liaisons. Preserving Post-War Modernism in City Centres*, Anja Kervanto Nevanlinna (ed.) (Vaasa, Tkkösoffset Oy, 2002).

Chapter 8

Planning the Historic City:
1960s Plans for Bath and York[1]

John Pendlebury

John Pendlebury is a Senior Lecturer within the Global Urban Research Unit, of the School of Architecture, Planning and Landscape, University of Newcastle (England). This chapter forms part of on-going research about the management of British historic cities in the twentieth century. In 2005 he was awarded a grant from the Arts and Humanities Research Council for the cataloguing, conservation and management of personal papers of Thomas Sharp, housed at Newcastle University.

Introduction

Issues of how to balance planned modernity with the conservation of the character of historic city came to the fore across much of the developed world in the 1960s. In Britain, this was the decade that saw the first explicit legislative recognition of the importance of historic areas through the creation of 'conservation areas' in the 1967 Civic Amenities Act, following a flurry of activity by government and others on appropriate ways to plan the historic town. A huge range of work dealt with these issues in this period including international exhortations,[2] major conferences,[3] official statements[4] and coverage in other key documents of the period.[5]

---

1   A longer version of this chapter, more fully considering the 1940s plans discussed, can be found at John Pendlebury, 'The Modern Historic City: Evolving Ideas in Mid-20th-century Britain', *Journal of Urban Design*, 10/2 (2005): pp. 253-273.

2   Council of Europe, *The preservation and development of ancient buildings and historical or artistic sites* (Council of Europe, 1963).

3   P. Ward (ed.), *Conservation and Development in Historic Towns and Cities* (Newcastle upon Tyne: Oriel Press Ltd, 1968).

4   Ministry of Housing and Local Government, *Preservation and change* (London: HMSO, 1967).

5   C. Buchanan, G. Cooper, et al., *Traffic in Towns: A study of the long term problems of traffic in urban areas. Reports of the Steering Group and Working Group appointed by the Minister of Transport* (London: HMSO, 1963).

Applied at the local level these concerns famously led to the four studies commissioned in 1966 for Bath,[6] Chester,[7] Chichester[8] and York,[9] each jointly commissioned by the government and the relevant local authority. As well as providing lessons for the individual cities their purpose was to inform more widely. The idea of the studies emerged from a conference in January 1966 convened by the Minister of Housing and Local Government, Richard Crossman. Originally intended to be five in number the fifth town, King's Lynn, dropped out as it was unwilling to meet its payments. Bath, Chester and Chichester were all apparently reasonably keen, although Chichester insisted that the work was done in-house: York wavered until York Civic Trust volunteered to meet half the local contribution and the local authority also sought to circumscribe the terms of reference.[10] Yet part of the logic of the selection of these cities was that they were more active than most in area conservation.[11] In Bath Colin Buchanan was chosen having undertaken a range of commissions in the city in the years immediately preceding.[12] In Chichester the study was undertaken by the County Planning Officer and followed the development plan which had recently been reviewed[13]. There was much pressure in York from the Civic Trust and other bodies for a more co-ordinated approach to conservation and a 'Town scheme' of grant assistance had been commenced in 1964,[14] and Chester was one of the few authorities that used powers introduced in 1962 to make grants to conservation works from money raised locally through the rates.[15] These four studies of historic towns remain as often cited benchmarks in the development of thought about appropriate responses to the planning of historic towns.[16]

Prior to this period planners often saw conservation and preservation activity as very much at the fringes of mainstream planning. Yet there is an earlier group

---

6   Colin Buchanan and Partners, *Bath: A Study in Conservation: report to the Ministry of Housing and Local Government and Bath City Council* (London: HMSO, 1968).

7   Donald Insall and Associates, *Chester: A study in conservation: report to the Ministry of Housing and Local Government and Chester City Council* (London: HMSO, 1968).

8   G.S. Burrows, *Chichester: A study in conservation: report to the Ministry of Housing and Local Government and Chichester Borough Council* (London: HMSO, 1968).

9   L. Esher, *York: A study in conservation: report to the Ministry of Housing and Local Government and York City Council* (London: HMSO, 1968).

10   W. Kennet, *Preservation* (London: Temple Smith, 1972).

11   D.M. Palliser, 'Preserving our heritage: the historic city of York', in R. Kimber and J.J. Richardson (eds), *Campaigning for the environment* (London: Routledge and Kegan Paul, 1974) pp. 6–26); and John Gold, 'York: A Suitable Case for Conservation', *Twentieth Century Architecture: The Journal of the Twentieth Century Society*, 7 (2004): pp. 89–100.

12   For example, Colin Buchanan and Partners, *Bath: A Planning and Transport Study* (London: Colin Buchanan and Partners, 1965).

13   West Sussex County Council, *Preservation and Progress* (Chichester: WSCC, 1965).

14   J. Shannon, *York Civic Trust – The First Fifty Years: Preserving, Restoring, Enhancing, Enriching England's Second City* (York: York Civic Trust, 1996).

15   W. Kennet, *Preservation*.

16   J. Delafons, *Politics and Preservation* (London: E and FN Spon, 1997).

of plans that collectively form a major body of work on the nature of planning for historic towns and cities. During the course of World War II and in its immediate aftermath a whole series of plans (now collectively referred to as 'reconstruction plans') was produced for a wide spectrum of settlements in the UK. Not surprisingly, badly war-damaged cities usually commissioned plans, but many were produced for settlements untouched by bombing and many for historic cities. These vary enormously in terms of approach and sophistication in their attempts to reconcile the historic qualities of place with functional modernity.[17]

The principal focus of this chapter is the 1960s government-sponsored conservation plans for two historic cities, Bath and York, but these are briefly placed in the context provided by their 1940s predecessors. Attention is mainly directed to how the plans conceived the historic city, rather than, for example, how much the plans were implemented, although there is some brief discussion about how they were received and their legacies to the cities concerned.

Bath

Buchanan's 1968 report[18] was a culmination of his work in the city and needs to be understood in relation to his other work in Bath. In particular he was the author of *A Planning and Transport Study* commissioned by the Council in 1965,[19] following an earlier report on traffic in Bath in 1964. Traffic was the central preoccupation of the 1965 report. There was seen to be a terrible dilemma between relieving the city and its heritage of traffic and finding routes to achieve this that did not impact upon the heritage. The key elements of the historic environment were defined highly selectively. Broadly the focus was upon the medieval city core and the show-piece elements of the Georgian town, including the Circus and Royal Crescent. Extensive comprehensive redevelopment was proposed, or at least accepted, for most of the city centre to the south and west of the historic core and for more limited areas elsewhere. Vertical segregation of traffic and pedestrians was regarded as desirable for the larger redevelopment areas.

Different options were proposed to solve the traffic problem. All contained a tunnel to carry traffic from west to east in the city, extended as a cutting through the residential area of Bathwick. There was also a proposed central cross-route in a 20-foot cutting, close to some of the Georgian set-piece developments.

---

17   P.J. Larkham, 'The place of urban conservation in the UK reconstruction plans of 1942–1952', *Planning Perspectives*, 18 (2003): pp. 295-324; John Pendlebury, 'Planning the Historic City: 1940s Reconstruction Plans in Britain', *Town Planning Review*, 74/4 (2004): pp. 371-393; and John Pendlebury, 'Reconciling history with modernity: 1940s plans for Durham and Warwick.' *Environment and Planning B: Planning and Design*, 31/3 (2004): pp. 331–348.

18   Colin Buchanan and Partners, *Bath: A Study in Conservation.*

19   Colin Buchanan and Partners, *Bath: A Planning and Transport Study.*

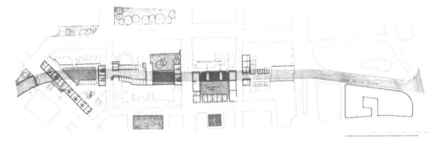

Figure 8.1   Buchanan's cut-route

Sources: Illustration is from the 1965 report (Colin Buchanan and Partners, *Bath: A Planning and Transport Stud.* (London: Colin Buchanan and Partners, 1965)); the plan is the modified proposal from the 1968 conservation study (Colin Buchanan and Partners, *Bath: A Study in Conservation: report to the Ministry of Housing and Local Government and Bath City Council* (London: HMSO, 1968)).

The more drastic traffic possibilities, not favoured by Buchanan, included a four-lane riverside route, skirting the historic core. Car parking, including for the core Georgian area, was to be provided through the construction of multi-storey car-parks.

The subsequent 1968 conservation study[20] was limited in that it only dealt with a study area, defined by the City Council, of the part of the city that had been occupied by the medieval town. Thus, it did not include the Georgian expansion for which Bath is principally famous. The 1968 plans were demonstration studies of the practicality of reconciling preservation objectives with modern functionality and in the case of Bath the complexity of the central area was considered more representative of such problems than the Georgian set-pieces.

The plan emphasised Bath's importance, 'in an English context Bath is one of the half dozen most precious small towns, for its architectural quality, for its historic associations and its contribution to the art of urban design'.[21] Greatest stress was placed upon the City's visual qualities, for 'in Bath as a whole the façades are very much more important than the interiors'.[22] The existing listing of buildings was considered to under-represent group value and the role of buildings as part of visual compositions. Conversely de-listings were suggested, including some altered works by John Wood Senior, one of the key architects of Georgian expansion.

The study area was divided into four and ranked as highest importance, secondary importance, little importance and in need of large-scale renewal. With the first category preservation was held to be imperative. With the secondary areas the aim was to keep the best buildings and to conserve the general character, though 'large-scale renewal cannot be ruled out'.[23] In the other areas the policy 'must be the acceptance of change'.[24] Areas one and two covered approximately 60% of the area. Thus Buchanan was advocating major change to at least 40% of the historic core, and possibly more. The inner road, previously in a cutting, was now covered in a tunnel. As proposed it involved demolition of historically significant buildings in Old Bond Street and Queen Street. Rebuilding to create general character rather than precise architectural form was recommended for Queen Street but in Old Bond Street the overall composition was to be re-established. There was a general strong recommendation that new build should not be neo-Georgian, except where completing a unified composition.

Thus then, traffic was a continuing preoccupation, including in the medieval core of the city, and whilst there was some discussion and study of traffic management, major road construction was deemed to be necessary. The other key issue was held to be finding new uses for historic buildings. This was considered to be difficult but achievable depending upon major resources, imagination and determination, with

---

20 Colin Buchanan and Partners, *Bath: A Study in Conservation.*

21 Ibid., p. 10.

22 Ibid., p. 13.

23 Ibid., p. 47.

24 Ibid., p. 48.

the benefit of Bath being a university city with a consequent demand for flats that students might occupy.

The successive proposals for conserving Bath by Buchanan were strongly modernist in character. Though the significance of place was clearly articulated and preservation a key objective, it was a highly selective approach principally based around architectural quality and picturesque effect. Preservation was intended to sit alongside massive transformation and this was seen as compatible with sustaining the historic character of the city.

This vision was soon to be subject to a very public critique. Perhaps in Bath more than any other British city was there a fierce backlash in the early 1970s over how the historic city was being planned and managed. For example, Bath featured prominently in pro-conservation polemics of the period[25] as well as generating at least two texts specifically on the perceived destruction of historic Bath[26] and scathing comment on planning in the city in the professional press.[27] These were not necessarily aimed specifically at Buchanan's influence in Bath. Fergusson[28] directed most of his ire at the local authority and some of their other architectural advisors. He noted the constraints Buchanan was placed under by the briefs he received from the city and the pre-existing Development Plan and the way that some of his recommendations had been ignored. However, he did also note a letter by Buchanan to *The Times* in 1972 which seemed puzzled by the contestation over the redevelopment of 'minor' Georgian architecture and to implicitly support the local authority's approach. Central to the critiques was the idea that the more modest Georgian heritage, 'artisan Bath', had been undervalued, 'Every attack on a minor Georgian building is an attack on the architectural unity of Bath'[29]. Artisan Georgian buildings were argued to be important as part of the story of Georgian Bath, because of their spatial role in linking the grand compositions, and high quality, serviceable buildings in their own right. To the critics the significance of Bath was believed to lie in its totality as an artefact, principally of the eighteenth and nineteenth centuries, rather than in the architectural set-pieces as such. This was compounded by replacement modern constructions that were at best indifferent. The tunnel proposal was particularly objected to (the Bath Corporation's scheme was even more interventionist that Buchanan had proposed) and it was finally abandoned in 1976.

The earlier 1945 plan for Bath was produced by Patrick Abercrombie in co-authorship with the City Engineer (Owens) and the Planning Officer for the Joint Area Planning Committee (Mealand)[30]. The details of the approaches taken to

25   T. Aldous, *Goodbye Britain?* (London: Sidwick and Jackson, 1975); and C. Amery and D. Cruikshank, *The Rape of Britain* (London: P. Elek, 1975).

26   P. Coard and R. Coard, *Vanishing Bath* (Bath: Kingsmead Press, 1973); and Fergusson, A., *The Sack of Bath* (Salisbury: Compton Russell, 1973).

27   Architectural Review, 'Bath: City in Extremis', *Architectural Review*, 153/915 (1973): pp. 280–306.

28   Fergusson, *The Sack of Bath*.

29   *Architectural Review*, 'Bath: City in Extremis', p. 280.

30   S.P. Abercrombie, J. Owens, et al., *A Plan for Bath: The Report Prepared for the Bath and District Joint Planning Committee*. (Bath: Bath and District Joint Planning Committee, 1945).

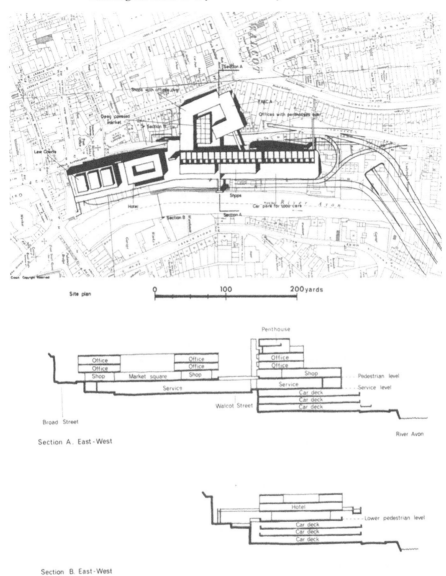

Figure 8.2   Walcot Street area proposals by Colin Buchanan and Partners
Source: Colin Buchanan and Partners, *Bath: A Planning and Transport Study*, (London: Colin Buchanan and Partners, 1965).

Figure 8.3   Part of the Walcot Street area as built

planning Bath by Abercrombie and Buchanan differed. Planning concepts used by Abercrombie were developed and refined by Buchanan, so Buchanan deployed the environmental area concept he had developed, rather than the land-use precincts favoured by Abercrombie. Buchanan viewed environmental areas as urban 'rooms', areas not free from traffic, but free from extraneous traffic and where environmental quality predominated over the use of vehicles.[31] The detailed nature of proposals also changed. For example, both plans included proposals for comprehensive redevelopment in the Walcot Street area. Abercrombie's proposal was illustrated with neo-classical buildings, Buchanan's with vertically segregated brutalism.[32]

However, fundamentally there was a great deal of continuity of approach and underlying principles. For example, both plans had a real sensitivity and understanding of the significance of Bath whilst ultimately only considering it vital to retain the major ensembles of Georgian development. Both ultimately saw the potential and need for major redevelopment and change, especially in the south and east of the centre, including the loss of significant amounts of Georgian building. Both acknowledged the environmental problems of traffic but saw major road building as inevitable. Both, despite their sensitivity to place, were ultimately technocratic and sought to modernise by introducing a very clear twentieth century stamp to substantial areas of the city. This was a way of viewing Bath discredited in the 1970s and today the city is a World Heritage Site with an emphasis firmly on

---

31 Buchanan, Cooper, et al., *Traffic in Towns: A study of the long term problems of traffic in urban areas. Reports of the Steering Group and Working Group appointed by the Minister of Transport.*

32 Bath City Council, *Forty Years On* (Bath: Bath City Council, 1985).

sustaining historic character across the city. Rightly or wrongly, many of the more recent developments have assumed a neo-classical character.

York

The starting point for the York study was not auspicious. The appointed consultant Lord Esher (also know as Lionel Brett) was apparently a reluctant participant[33] and he was not warmly welcomed to York: 'I was smuggled into the Guildhall by a back door for fear I might meet the Press, then told by Mr Burke (railwayman and leader of the Labour Council), in his solid Yorkshire drawl, "we don't like consultants here". The Conservative boss was if anything more unfriendly.'[34]

His study[35] set out five objectives for the City:

1. That the commercial heart of York should remain alive and able to compete on level terms with its neighbour cities, new or old.
2. That the environment should be so improved by the elimination of decay, congestion and noise that the centre will become highly attractive as to a place to live in for families, for students and single persons, and for the retired.
3. That land uses which conflict with these purposes should be progressively removed from the walled city.
4. That the historic character of York should be so enhanced and the best of its buildings of all ages so secured that they become economically self-conserving.
5. That within the walled city the erection of new buildings of anything but the highest architectural standard should cease.

The report started with a sophisticated and lyrical analysis of the character of the city. For example, in describing the form of the city, 'The streets themselves are often no more than slits in the dense texture of buildings, or alleys running off under low openings to dwellings and workshops giving on to tiny yards'.[36] This townscape analysis related quite closely to the qualities identified a few years earlier for York by Pace.[37] Emphasis was placed on the richness of the propinquity of buildings of different styles and periods in forming the character of the city.

A key element of the proposals was to make the walled city liveable with an emphasis firmly on increasing the number of people living within the Walls. At the time of the study the residential population was 3,500, a figure that represented a long decline as commercial and industrial development had displaced people from

33   Gold, 'York: A Suitable Case for Conservation'.

34   L. Brett, *Our Selves Unknown* (London: Victor Gollancz, 1985), p. 163.

35   Esher, *York: a study in conservation*.

36   Ibid., p.15.

37   G. Pace, 'The York Aesthetic', in *Annual Report of York Civic Trust* (York: Civic Trust, 1961/62).

Figure 8.4   **An extract from Esher's proposals plan. The part of the city shown is the commercial centre. Interventions proposed are shaded and show the proposed narrowing of the nineteenth-century Parliament Street.**

Source: L. Esher, *York: a study in conservation: report to the Ministry of Housing and Local Government and York City Council* (London: HMSO, 1968).

the centre from the nineteenth century. The target was to increase this to 6,000. Measures proposed to achieve this included first, removal of some (but by no means all) industry from the walled city, with a focus on relocating industrial uses which generated significant traffic or that made a noxious neighbour. Some comprehensive redevelopment was also envisaged, but focused on low-grade industrial use and was very modest and surgical when compared with proposals for Bath. Second, the potential for the re-use of vacant upper floors was emphasised, and there was a study of the Petergate area specifically on this issue.

Another key theme for making the city more liveable was the management of traffic, including some pedestrianisation, although greater emphasis was placed upon restricting access to vehicles to the historic core more generally through the use of permits. Car parking was to be principally provided by four multi-storey car-parks, two within the Walls and two without. The most brutal in its impact would have been a car-park spanning Piccadilly, on the south side of the centre. Novel means of achieving local circulation were proposed with electric vehicles suggested for bringing shoppers in from the multi-storey car-parks and for making deliveries during the day. Other management proposals for the walled city (very little by way of new road construction was proposed) included the prevention of cross-town traffic, and the *narrowing* of one of the major nineteenth century streets, Parliament Street.

Esher's terms of reference precluded him from considering traffic proposals for the city as a whole. However, he was very sceptical about the inner ring road that was being proposed immediately outside his study area, just beyond the Walls. His objections were made both on amenity and functional grounds. It was considered that such a road and its associated works, such as roundabouts and junctions, would dwarf the City Walls and 'above all the Bars, whose impressiveness is dependant on the contrasting scale of the small buildings in their vicinity'.[38]

In addition to the focus on liveability, Esher considered that York had nowhere near reached its potential as a tourist city and that there was some potential for commercial development. However, the report was very sensitive to the issue of building mass and large multiple stores were to be only allowed in limited locations. Other elements of the study included a re-evaluation of the listing in the city, with the purpose of adding many buildings of townscape value. Though it was acknowledged that some of these would in turn be lost as part of the natural evolution of the city, it was indicative of a more inclusive approach to what was considered worthy of protection and retention. Furthermore, though it was made clear that this was a planning study it was also acknowledged that the contribution of historic buildings was not essentially visual, that fabric was significant too. Overall, though the report was not without its solecisms, such as the brutal multi-storey car-park proposed for Piccadilly, there was a feel of a fine-grained sensitivity to the City. It feels like a study worked out from street-level and indeed has lots of photographs of people animating space, rather than technocratically using plans and models.

---

38 Esher, *York: a study in conservation*, p. 53. 'Bars' are gates into the walled city.

**Figure 8.5   A photograph from Esher's York. Like many of the photographs
in the report it shows places and people.**

Source: L. Esher, *York: a study in conservation: report to the Ministry of Housing and Local
Government and York City Council* (London: HMSO, 1968).

Figure 8.6   Housing in Aldwark

Esher's study generated much publicity[39] and was generally very well received in the professional press[40] and in York itself[41] and continued to be a touchstone in the city for many years[42] and indeed was still being seen as a significant reference point by developers, the local authority and objectors at the public inquiry over controversial proposals to extend the Coppergate shopping area in 2002. Although the Council's

39   L. Esher, *Esher's York: A Study in Conservation* (York: Yorkshire Evening Press, 1969).

40   Gold, 'York: A Suitable Case for Conservation'.

41   Palliser, 'Preserving our heritage: the historic city of York'; N. Lichfield and A. Proudlove, *Conservation and Traffic: A Case Study of York* (York: Ebor Press, 1976); and P. Nuttgens, *York: The Continuing City* (London: Faber and Faber, 1976).

42   Shannon, *York Civic Trust – The First Fifty Years: Preserving, Restoring, Enhancing, Enriching England's Second City.*

report on the Esher proposals was described at the time as 'ill informed, misleading and tendentious'[43] the local authority ultimately accepted most of the proposals (with reservations about costs) and Esher was subsequently engaged to help implement the redevelopment of the Aldwark area, one of the major areas of new housing within the Walls.

However, the Council pushed ahead with proposals for an inner ring road, which were bitterly contested. New amenity groups prepared to challenge the authorities more directly were formed and the road proposals were ultimately defeated at public inquiry.[44]

The earlier reconstruction plan for York was commissioned from a leading consultant, S.D. Adshead in 1943, though he was dead by the time the report was completed and subsequently published.[45] The two plans for York did show some continuity. For example, both stressed the significance of narrow streets in the historic centre to the character of the City and the need to make these a tolerable environment for the pedestrian through, for example, pedestrianisation. However, what is more striking is how in other ways there was a fundamental shift. In particular, the degree of intervention seen to be necessary and desirable for redevelopment and for accommodating traffic was vastly scaled down. The view in the Esher report was that massive road proposals in or adjacent to the walled city would inevitably be to the detriment of the character of the city and therefore other means of managing traffic must be used. Essentially with the Adshead plan it was a case of identifying buildings and areas of particular quality and redeveloping much of the rest; with Esher of keeping the fabric of the city unless there was a particular need to redevelop. With Adshead the overall result would have been something of a de-intensification of the use of the historic core, with Esher the intention was to intensify the residential use of the area in particular.

Sea Change or Evolution?

The four 1960s conservation studies have often been considered one of the key defining moments in a sea-change of approach towards conservation and redevelopment in the late 1960s and early-1970s. Although it is certainly true that profound changes occurred at this time, in terms of the retreat of a modernist doctrine of comprehensive planning and the rise of a conservationist agenda, it is the contention of this chapter that the studies considered here, for Bath and York, represent degrees of evolution

---

43 Anonymous, 'Conservation International', *Architectural Review*, 146/874 (1969): pp. 409–410. p. 409.

44 D. Cummin (ed.), *York 2000: People in protest* (York: York 2000, 1973); Palliser, 'Preserving our heritage: the historic city of York'; and Lichfield and Proudlove, *Conservation and Traffic: A Case Study of York.*

45 S.D. Adshead, C. J. Minter, et al., *York: A Plan for its Progress and Preservation* (York: York Corporation, 1948).

of approach to historic cites, important antecedents for which can be found in earlier plans for the cities of the 1940s.

The 1968 plans and their predecessors were all essentially masterplans in basic approach, and all were grappling with problems of reconciling the historic city with modernity. All advocated a balance between the conservation of historic character with the continuing evolution of the living city, albeit the balance suggested between plans varied significantly. Provision for traffic was a major issue in both cases, as was the need for pedestrians to be able to enjoy historic areas free of excessive numbers of vehicles. The need for cities to be functioning modern places and the need to rationalise land-use to some degree was a common issue, but at the same time, though the emphasis varied between plans, there was a recognition of the richness of mixing different uses considered to be compatible neighbours. The need to find new uses for historic buildings whose original or existing uses were obsolescent was another recurrent theme. Generally plans for both periods had a clear ideology that new buildings should be clearly contemporary in style. The Abercrombie plan for Bath was unusual in allowing for the possibility of new buildings closely following historic precedent, possibly because of the historic unity of building style in the city.[46]

Contemporary reviews of the 1968 plans tended to focus upon government inertia in implementing the plans' recommendations[47] and to see the group of four plans as similar in approach, for example Hall[48] considered that they all had a rather similar civic design focus. However, with hindsight it is possible to see that the Buchanan and Esher plans had some profound differences in how they sought to manage the historic city. They shared some similarities of outlook. For example, there was the same rhetoric about the need for progressive planning and the same distaste for pastiche or historicist architecture. Modish 1960s solutions are evident in both. Buchanan advocated vertical segregation in Bath and the Esher plan contained brutal multi-storey car-parks. However, there was a selectivity in identifying the historic city in Bath reminiscent of the 1940s plans, whereas in York the historic city was conceived more inclusively and extensively and as a more intricate series of intimate visual relationships. Approaches to dealing with traffic were also quite different between the two plans. Buchanan was, of course, the most famous British writer on traffic in the 1960s. In the *Buchanan Report*[49] and in earlier writings[50] he had displayed a deep ambivalence to the rise of the motor car and its impact on urban life; he believed roads to be necessary but recognised their destructive qualities. He

---

46 Pendlebury, 'Planning the Historic City: 1940s Reconstruction Plans in Britain'.

47 Architectural Review, 'Dropping the Pilot?', *Architectural Review*, 148/886 (1970): 340–348.

48 P. Hall, 'Salvaging Our Historic Towns', *New Society* 349 (1969): pp. 872–4.

49 Buchanan, Cooper, et al., *Traffic in Towns: A study of the long term problems of traffic in urban areas. Reports of the Steering Group and Working Group appointed by the Minister of Transport.*

50 Colin Buchanan, *Mixed Blessing: The Motor In Britain*. (London: Leonard Hill, 1958).

saw the need to contain the use of road traffic but also the need for extensive new urban roads and it was this balancing act between management and provision he sought to achieve in his work in Bath, seeking to minimise and mitigate the impact from what would inevitably be a major intervention in the urban environment. The City of York was also proposing major new urban roads, but Esher challenged this in terms of both the aesthetic impact a ring road would have had as well as its functional logic. He was signalling an approach with a greater emphasis on the management and containment of traffic. It is tempting to ascribe the differences in approach in part to the professional backgrounds of the consultants, Buchanan an engineer used to solving difficult technical problems vs. the aristocratic sensibility to tradition of the architect Esher.

The Buchanan studies of Bath did not represent a significantly more pro-conservation approach than their 1940s precursor; indeed, in some respects, such as the scale of roads proposed and vertical segregation and brutalist architecture, their proposed interventions were more drastic. However, with Esher's study of York we can see a distinct step-change; most obviously through the relatively small level of proposed redevelopment, the emphasis on intensifying use and especially residential use within the Walls and a changed emphasis upon the relationship with the car and providing for its use. This latter became the key environmental battleground of urban areas in the 1970s and was of great significance in both cities considered here, with, in both cases, ambitious road proposals abandoned as the result of a loss of faith in such solutions to traffic issues and the realisation of the full costs of such schemes in environmental and financial terms.

The Esher plan was to prove particularly influential; nearly forty years later Esher is still considered a major benchmark in conservation-planning in York, whereas in Bath it is the reaction to modernist planning, *The Sack of Bath*,[51] that is the touchstone.[52] Esher was part of a wave of activity that helped people evolve a new way of seeing the historic city which advocated drastically reducing the level of proposed planned intervention. What is remarkable looking at the Esher study today is how, in most respects, fresh and contemporary it feels with its focus on creating a liveable city. As Gold[53] has written, 'It was the spirit of urban sustainability without the contemporary lexicon.'

----

51 Fergusson, *The Sack of Bath.*

52 A. Harrap (Bath World Heritage Site Officer), Interview with John Pendlebury 2/7/2004.

53 Gold, 'York: A Suitable Case for Conservation', p. 98.

Chapter 9

Multiple Exposures or New Cultural Values? European Historical Centres and Recent Immigration Fluxes

Alessandro Scarnato

Alessandro Scarnato graduated at the School of Architecture of Florence (Italy) and is currently (2005) finishing his Phd programme at the Universitat Politecnica de Catalunya in Barcelona, Spain. As an architect, he works in the fields of restoration and public space, and has been awarded in international competitions. He collaborates with Italian magazines and teaches in private institutions of design such as the IED.

The importance of historical and cultural values has increased considerably in the projects of renewal or revitalisation of European centres in recent years. More and more, the concept of 'identity' is used as a reference by planners and politicians, raising it to the same level of importance as other topics such as the transportation network, the hygiene requirements or the structural conditions of the buildings. Furthermore identity, in most cases, is also an essential element for the so-called 'urban marketing'. Many local administrations work carefully on their image (or *identity*) in order to attract tourism, economical funding from central or international institutions, and organisation of sport or public events.

The role of the historical and monumental centres, in this construction-of-identity process, is evident as well as the fact that the traditional residents are no longer the only ones living in and using these areas. Other new communities (most of them extra-European) are often settled in these historic centres. As a consequence, the heart of European towns – the core of local identity, scene of historical episodes and seat for the most important institutions – is then altered by the intense presence of several kinds of immigrants to these areas. We will later analyse the reasons for this choice, sometimes an effect of economic conditions and sometimes the act of settlement inside a specific urban environment. In any case, these arrivals provoke the re-introduction of lost uses of social spaces, while others acquire a new identity. Is this a merging of old values inherent in these centres together with new values brought by new inhabitants? or is it a progressive substitution that will inevitably erase the identity of European cities? Is it rather a mere co-presence of two images of the same place, used and felt in different ways by different actors, that will produce, in the long run, the confusing effect of multiple exposures on the same frame? We concentrate our attention on historic centres because of their importance for the identity of many European towns and for the different panorama compared to the peripheral areas, already studied by anthropologists, sociologists and urban planners and efficaciously represented in many works by writers, film directors and songwriters.[1]

This is only a first attempt to approach this subject matter, urgent for the great effects that it can have in a political sense, but not studied sufficiently by architects and planners. First, we will analyse the main features of these new fluxes of immigrants. Later we will talk about the actual use of public spaces in old centres of southern Europe, analysing the most significant changes in the way in which these spaces are inhabited and used. Two towns will be taken as examples of the dynamics involved in the transformation of southern European historic centres: 1) Barcelona, Spain, a city where the historical centre lay in strong degradation for more than a century; and 2) Florence, Italy, as the prototype of the European City of Art.

The final part will be a reflection about how immigration can change or influence the construction of identity in European historic centres, taking into account that, in such a vivid subject, to make the right questions happens to be as, or even more, important as to have the answers.

Why the New Immigration is New?

Immigration is not a recent phenomenon in Europe. Throughout history, the mass movement of populations from one area to another has always been present and has influenced the development of the national identities. However, over the last twenty years the reasons for immigration has changed. It is the desire to improve the quality of life that inspires immigrations of today. Due to great technological advances, and a new European political scene, immigration has increased and has touched in a noticeable way historical town centres. An analysis of these points will be useful in order to understand the peculiar essence of these 'new immigrants'.

The new immigrations are not strictly connected to substandard political and/or economical conditions. These immigrants are not necessarily refugees, which should be considered in the main as an extreme condition. The majority of them are looking for a change in their life, an improvement that is seen as difficult or impossible in their homeland.[2] From this point of view, it is possible to argue that a political or economic

---

1   A list of the artists who describe life in peripheral areas would be too large. Here we can just mention the work of Italian director Pier Paolo Pasolini (1922–1975) and, in recent years, British director Ken Loach. The French movie *La Haine* (The hate, 1995) by director Mathieu Kassovitz is another important essay about the difficult relationship among different cultural and social environments in modern peripheries.

2   It is necessary to explain what we mean talking about 'immigrants': a first group, the most consistent, are the temporary ones. They come without their families planning to make

crisis is seen, in many cases, only as an accelerating element for a decision that was made or conceived a long time before. Furthermore, these new immigrants do not arrive in Europe unprepared, but follow an already structured three-step process: 1) at home, where they make the decision to leave and choose which members of the family will be immigrating and for how long; 2) the actual journey which can change depending on political situations and the cost of transportation (in the last five years, the north-African routes have become the most covered and the Strait of Gibraltar is now the main gateway of entry to Continental Europe; and 3) the last step is the most complicated: arrival and the actual integration process.

It is important to emphasise that the people coming from countries far from Europe are not escaping from historical traumas. They are often in search of a higher standard that will allow them the possibility to return to their homeland in a better financial situation. These immigrants have the 'luxury' to individualise their new environment, reflecting their personal cultures, as opposed to refugees. Moreover, newcomers set up their own commercial premises as soon as possible or try to enter in specific sectors like construction.

Almost all immigrants move following an established route and very rarely travel on their own. They move only if they have enough money to pay for the journey.[3] Individuals with official identification, can now easily afford intercontinental flights thanks to low-cost airlines. At the same time, the Internet and the low cost of telecommunications allow people to leave the place of home without feeling forced to make a definitive break. These same technologies also simplify the advanced contacts to persons in their ethnic group in the new location making it easier to be absorbed by the new community.

As a consequence of this, European towns have changed very quickly because of the newcomers, very often being people without any kind of previous relationship to their new countries: Pakistanis in Spain, Chinese in Italy, or Philippines in France. This demographic change within the historical centres of towns and cities has been highlighted by the clear visibility that immigrants have in the city; not just

---

money abroad for a number of years; then, depending on the evolution of the personal situation, they can come back or they take the rest of the family to the new hometown. Then there are individual immigrants, who don't consider coming back as a priority and that normally make more attempts towards an effective integration in the new place. Finally, there are the families: from the rich North American family who bought a flat abroad to spend the holidays, to the big East Asian family moving entirely with all its members. This last group doesn't make many efforts to integrate themselves into the new country, and they rather close their communities, many times with a clear, recognisable visual identity. Source: *La presenza straniera nell'area fiorentina* (Firenze, 2003) and: M. Baldisseri, 'In casa d'altri: spazi sociali di donne peruviane a Firenze', in P. Clemente and M. Loda, M. (eds), *Migrare a Firenze* (Firenze: Comune di Firenze, 2003).

3   The horrible images of hundreds of immigrants suffering in abandoned boats or containers move one to mercy, but the dramatic conditions of these journeys are mostly due to the attempt to elude controls at the boarders, rather than to represent the unique affordable way to move.

because of their appearance, but because of their intense use of streets, public spaces, transportation and services.

Barcelona, the Centre as 'Work in Progress'

Barcelona is a modern metropolis with a 2000-year-old core, still alive in the symbolic and social sense, a core that was changed considerably when the famous planner Ildefons Cerdà designed the expansion on the surrounding plain in 1859. The rich bourgeois preferred to move to the new districts and the Ciutat Vella (Old Town) started an unstoppable decaying process. The poor and the immigrants to Barcelona quickly became the first demographic presence. In large sectors of the centre, historical monuments remained as islands surrounded by a sea of misery, poverty and sickness. Projects of renewal for Ciutat Vella became a constant topic in political debates, but too many uncompleted urban proposals over decades only amplified the abandonment of this part of the town.

When the Ajuntament (the local city council) finally began an effective restoration of the district in the mid 1980s, the historic centre of Barcelona was irreparably turned into an urban black hole. Drugs, prostitution, violence and severe social problems were concentrated in a physical environment comprising decrepit buildings and abandoned monuments,[4] and narrow, dirty streets where a substantial part of the daily activities were illegal. In the middle of this environment, the ancient hill where the Cathedral, the City Council and the Regional Government of Generalitat are located, was the only part of the district where citizens and tourists ever went or visited. Most citizens of Barcelona never dared to go into many areas of the historic district, and this situation is perceptible still today if we notice the continuous highlighting of the Administration propaganda on the fact that now it is possible to enter and to live in Ciutat Vella.

From 1975 to 2000, Barcelona has been touched by a process of an increasing political, cultural and tourism importance.[5] The high point of this process was the assignation to host the Olympic Games for 1992. Seven years later the city was

---

4   Even if some important buildings were still in a good state, a lot of monuments laid almost abandoned all over Ciutat Vella. Romanic churches such as Sant Pau del Camp, the aristocratic houses in the Ribera sector or singular pieces and decorative elements. This was due, probably, to the decision, in the early twentieth century, to create a 'Monumental district' where all the most significant buildings should be replaced after taking them down. In spite of this intention, the most part remained in the same place, but the fact that some was effectively moved covered with a veil of imminent destruction or removal the monuments of Ciutat Vella. Almost no restorations or even simple investigations were made for long years, and the interest for the possibility to rehabilitate the district as an entire urban tissue has always been very low.

5   The industrialisation of Barcelona entered in a strong crisis in the mid 1970s. It is not hazard to think that the cultural and political growth of the town in the following years can also be seen as a conscious reaction against the difficulties of the economical situation. For an extensive description of this process see: Francisco Javier Monclús, 'Barcelona's planning

Figure 9.1   Ciutat Vella, Barcelona

awarded the R.I.B.A. gold medal and definitely entered into the dimension of a European Capital City. These changes turned the city into a modern metropolis and an international architectural reference, an urban model where most of the solutions applied were considered the ultimate, above all concerning the peripheral sectors.

But the centre, as we said, was another story. Rehabilitation projects for Ciutat Vella took into account that the composition of the local population was made up of immigrants from other regions of Spain, above all from Andalusia, Murcia and Galicia, and a strong nucleus of natives from the district. Because of the common opinion that the bad living conditions were strictly connected with the physical situation of the district, it was no surprise that the old technique of massive 'esponjaments' (demolitions) was thought of as an essential rehabilitation instrument. The programme of interventions (opening up new squares, street widening, demolition and reconstruction of housing blocks) suffered because of several reasons: the urgency of the Olympic works, waiting for the economic support (provided by private enterprises and, above all, by the European Union Cohesion Funds), and the protests from inhabitants of the district, caused by the loss of a high percentage of

strategies: from 'Paris of the South' to the 'Capital of West Mediterranean', *GeoJournal*, 51 (2000): pp. 57–63.

historical urban tissue[6] and by the overall management of the allocation of residents, moved under expropriation.

Immigration from outside Europe erupted in the mid 1990s. Barcelona had already experienced massive immigrations, but always from Spain, and always after notable facts or changes in local political or economic situations. What happened in the 1990s was the arrival of a lot of people, mostly from the Maghreb countries, who were the first wave of many more in the following years. In the 1990s Ciutat Vella became the ideal place to live for a lot of immigrants who entered into the district because of the great number of abandoned flats waiting to be demolished or rented at very low prices. Furthermore, the district offered good opportunities to those who were looking for a safe place to hide, because of their illegal activities or their lack of legal residence documentation, especially in the two areas where the rehabilitation projects were more incisive: the Raval and Santa Caterina.[7]

A curious fact happened in the historic centre of Barcelona: a rehabilitation promised for decades, has been completed almost entirely after a 25-year process but the population that it was thought for has changed in the mean time. The City Council has had a lot of difficulties in understanding this process, maybe because of the high speed with which things happened. Ciutat Vella has now lost the poor and decrepit atmosphere that typified it for so long, but we may say that the transformation took place in a very different way, respectful to the original intention of the administration. The identity of the district is now a mix of secular testimonies of Catalonian history, for those who are able to read them, together with new ultra-modern episodes, such as Richard Meier's MACBA and exotic presences in the residential blocks.

Santa Caterina Market is, in this sense, extremely clear: the location of one of the most important Barcelona monasteries – until its destruction in 1836 – was the

---

6   According to official data, between 1988 and 2002, 2,147 buildings of Ciutat Vella have been rehabilitated (the 49%) while less than 500 have been destroyed in order to give to the district around 10 Ha of free public space. Sources: J. Casanovas (ed.), *Ciutat Vella, Ciutat construïda* (Barcelona: El Cep i la Nansa, 2003); J. Busquets et. al., *La ciutat vella de Barcelona, Un passat amb futur* (Barcelona: Ajuntament de Barcelona – Focivesa – UPC, 2003); and S. Von Heeeren, *La remodelación de Barcelona, Un análisis crítico del modelo Barcelona* (Barcelona: Veïns en Defensa de La Barcelona Vella, 2002). This last work presents an unusually critical point of view on the rehabilitation process of Ciutat Vella reporting the protests occurred for the way some demolitions took place, sometimes with a rude unconstraint regarding the actual state of conservation of the buildings and their historic, artistic or urban values.

7   The effects of such a big demographic change are difficult to estimate with official numbers. The local Departament d'Estadística de l'Ajuntament say that in early 1990s foreigner residents in Barcelona were around 25,000 (1.5 % of the total population in the town), while are now – February 2004 – almost 203,000 (12,8 % of tot. Pop.). A spectacular increase, above all for communities who multiplied their presence at amazing levels: Ecuadorians passed in less than eight years from 202 to 32,946; Moroccans from around 3,000 to almost 15,000; Pakistanis from 614 to more than 10,000 and they grow. More than 35,000 foreigners live in Ciutat Vella, mostly concentrated in the two sectors of Raval (Pakistanis) and Santa Caterina (Moroccans and Dominicans).

Figure 9.2   Barcelona

seat of the biggest market in the centre of city. When the rehabilitation of the district
started, the market had to change also and in 1996 the worldwide known architect
Enric Miralles was encharged to study a project in order to improve the commercial
building and the rest of Santa Caterina sector, seriously decayed after several decades

of degradation and the last ten years of inadequate or too aggressive rehabilitation policies. The market was originally supposed to open again at the end of 2001, but the discovery of archaeological remains stopped the works. In the meantime, the social and ethnical composition of the district changed as well. Approximately 35% of the previous shops preferred to close instead of waiting for a new market that will be finally inserted into a context characterised by a huge presence of Dominicans, Moroccans and Algerians.

In fact, all around the building the original Spanish or Catalonian shops have been progressively substituted by phone centres, Internet points, Muslim butcheries and other kinds of ethnic shops.[8]

Something similar happened in the sector of Raval, where the opening of a great new Rambla has not meant a new deserving habitat for original inhabitants. The urbanism of the *tabula rasa* provided, instead, the perfect environment for settlements of people that have found, in a vanishing part of the city, the opportunity to play an important role in the building of an urban identity that was no longer that of the previous inhabitants.

If the processes explained here represent an example of how a degraded centre can find a resource in a virtually conflictive presence like immigrants (due to the way they fill the voids and 'fix' the mistakes of wrong planning) another example is useful and stimulating, representing in many senses the opposite situation.

The Multiple Lives of the Centre of Florence

Florence, in Italy, has a marvellous centre famous for the magnificent beauty of its antiquities. It is the birthplace of the Renaissance and an authentic open-air museum, well conserved after traumatic rehabilitations in 1892 and 1936. In both cases popular areas were demolished for political and health reasons: in 1892 the old Middle Age market and the Jewish ghetto were sacrificed in the name of a more remunerative new image of Florence, in harmony with the aesthetic values promoted by contemporary urbanism. Forty-five years later, the Santa Croce district underwent a 'sventramento' (demolition) meant once again to remove hygiene problems, and also the social pressure of this area, particularly dangerous under Fascism. Since then, no other sensible urban change took place in the Centro Storico of Florence.

The city is now living promising years for city planning, but only in well determined and recognisable peripheral areas, linked to important infrastructures. Even if very few of the new projects are located in the centre, we can say that in this zone no important intervention was made in the last 50 years and the trend is clearly

---

8   Called Ethnic Commerce, the activity of selling every kind of goods, especially alimentary wares, in shops normally owned by people belonging to the same ethnicity of the customers. See: N. Monnet, *La formación del espacio público, Una mirada etnológica sobre el Casc Antic de Barcelona* (Madrid: Catarata, 2002).

conservative with a lot of protective restrictions for the historic tissue[9]. Florence is sold as a renaissance city where the most vivid experience that one could have is to walk in an artistic environment, where every detail or building reminds us of artists from the Middle Ages and the fifteenth century, the Quattrocento.

Such an attitude, promoted by many figures or institutions such as the City Council, central national organisms, and private companies, goes together with the real needs of the Centro, which suffers pollution, gentrification, high cost of living and a constant stressful use by thousands and thousands of tourists everyday. The most debated argument in local politics over the last 15 years has been car access restrictions to the centre. This is, even now, the hot-point for each administration joint with the slow but constant decrease of population and the closing of traditional shops.

Florence can be examined as a significant episode of the difficulties that a City of Art can have in finding a way to maintain both social and physical identity without destroying a consolidated image that attracts around 2.5 million tourists every year. Florentine population in the Centro[10] is decreasing in spite of the apparent vitality of this area that is really significant for every community, even when the people are living in another district.[11] Currently (in 2005) there is no existing scientific data about this fact, so we cannot be sure about the possibility that foreigners prefer the centre over something else because of its mere accessibility. Only immigrants from the so-called 'First World' (USA, Germany, UK or Switzerland) countries seem to stay in the historic district with the intention to feel surrounded by the 'Renaissance Air'.

Direct experience of people working in public institutions confirms that there is currently existing a kind of multi-level town in Florence. The presence of the communities of immigrants is hardly identifiable on the base of the place where they live, with the use of public space being the main indicator: the squares of Florence, no more accessible by car and turned into pedestrian areas, did not return to being typical piazzas for Florentines. On the contrary, we can characterise each square for the main foreigner presence. In Santa Maria Novella square, Somali women produce a colourful spectacle of traditional clothes every Thursday afternoon, when they meet. Peruvians and Albanians share the Station Square; Piazza Indipendenza is a constant and dense meeting point for Philippines; in Piazza Santo Spirito or in the

---

9   I. Massini and I. Saratti (eds.), *Parigi, Firenze, Barcellona, Recupero e valorizzazione del Patrimonio Architettonico: metodologie a confronto* (Paris: Association pour la recherche sur la ville et l'habitat, 2000), p. 125.

10   According to 2002 data from Istat, Italian National Statistics Institution, Florence has around 370,000 inhabitants. Almost 69,000 are in the centre, where are registered approximately 7,400 foreigners, distributed quite proportionally among different communities. Albanian, Philippine, Chinese, Moroccan, U.S.A. and Somali are the most numerous, but there is not a 'leading' presence compared to what happens in bigger towns.

11   The historic centre is the district with more registered foreigner residents, but in the five districts of the town there is a rather constant percentage respectful to the Italian population, around 2%.

Figure 9.3   Via Tornabuoni, Florence

historic Boboli Garden, South Americans are easy to meet, while the super-central Piazza Repubblica and its surroundings are the preferred place for North Americans and Europeans from the European Union.

It is possible to affirm that Florence does not have a clear territorial separation between these communities: there is rather a net structure of the Centro. Different communities are not divided into districts, and they use public spaces connecting a few focal points with streets where specific commerce can be found. The Cathedral, Piazza Repubblica, Piazza Strozzi and Ponte Vecchio are, in this sense, linked together by four streets, the most important of which is the elegant via Tornabuoni. This one has seen in recent years a strong process of unstoppable reverse-decadence due to the progressive substitution of almost every traditional shop with high fashion stores. Efforts to preserve the original fixtures of the historical shop locations have been unsuccessful: it is possible to visit a Max Mara store where the clothes are displayed on the old shelves of the previous centennial bookstore. Florentine people consider via Tornabuoni a sort of rich ghetto, where only Japanese tourists and the American or German residents can really enjoy going.

On the other hand, there are the streets for the other kind of immigrants. Via Palazzuolo, via della Scala and, above all, Borgo Panicale Street, close to San Lorenzo Market, in recent years have become authentic ethnic arteries where everything is somehow coming from, or dedicated to, people from the Third World.

History, People and Identity: Towards New Points of View to European Historic Centres

After looking at these two examples, it is easy to reflect on how city planning and architecture are involved in this new relation between physical historic European identity and new inhabitants. It is not a mystery that sociologists and anthropologists have been, until now, the most active in the effort to understand and manage the processes of demographic transformation, often seen as mere social problems by politicians and extraneous by architects and planners (with some exception, of course).[12] The ideal of a beautiful and efficient town for everybody has favoured a 'neutral urbanism' where public space was interpreted as an aesthetic and/or functional response to quantified needs. Even restorers and conservationists have not made many efforts in the research of the actual meaning of historic centres.

This is the problem. What do European historic districts and monuments really mean to contemporary Europeans? No doubt that in last 40 years, the use of centres has changed a lot, as well as the use of the rest of the city. But the spaces that have been the seat of local identity for centuries have suffered these mutations in a deeper sense: there's something perverse, indeed, in the decision to close with cages the porches of Renaissance churches in the centre of Florence, structures built expressively to recover the same homeless people that now are moved away. It is weird to notice that in Barcelona the rehabilitation process goes together with an intense effort to favour the birth of a new identity for the district, an identity supposed to be the release from misery of the past, but unfortunately not corresponding to the actual situation.

In more general terms, the way the historic public space is used has changed in a strong way after the 1970s: basically the function has changed. Public space is no longer seen as the seat of a specific task not possible at home, social interaction above all. The strong improvement of life conditions in modern apartments and the possibility to concentrate activities in huge commercial centres has turned public space into a simple 'outside' of the house, or into an unsafe place to experience only under precise conditions in specific hours. This process has become evident in the progressive elimination of filtering spaces between residence and public space: in historic buildings the existing filters (courtyards, entrance patios or stairways) are limited and sometimes their use is strictly forbidden.

A sensible discrepancy is then produced in old centres, where for centuries the communities have built their living landscape linking private houses together with public areas since everything was not affordable at home.[13] This was still taking place even at the beginning of the 1980s. The discrepancy lies in the gap between the

---

12 See the studies of Manuel Delgado and Kevin Lynch among the others. An important reference is also the work of Giandomenico Amendola, see: Amendola, G., *Uomine e case, I presupposti sociologici della progettazione architettonica* (Bari: Dedalo, 1984).

13 There is a large list of typologies of buildings and places that were supposed to be open to everybody: public baths, washtubs, stables and, of course, the most traditional ones like religious temples, markets, stages or weightings.

historic (and morphologic) identity and the actual identity of the urban environment. In other words, a contrast is currently going on when the native population talk about their old centres in Europe: there is a contradiction between the image they have of their centre and the real situation. Moreover, we can say that there is a conflict between the supposed values attributed to a certain space or symbol and the everyday meanings inherited by the current users.

As we have seen in Barcelona, the true problem of Ciutat Vella was not the physical decadence of the historic tissue (something solvable with a quite more delicate intervention) but the non-correspondence of the district to the image of a modern city that planners, influenced by the European examples of the early twentieth century, tried to impose on the Catalonian capital. The intervention was carried out in such a way that the previously mentioned processes of abandoning the social use of the town have been introduced and accelerated by the rehabilitation itself. Much of the original population decided to move, even if not forced through compulsory purchase procedures, on the basis of finding that outside of their houses, the city no longer provided the environmental quality capable of satisfying their basic needs, from social interaction to daily shopping.[14] The arrival of big communities of immigrants in a certain sense saved Ciutat Vella, even if the first wave were pretty conflictive. We do not think it a coincidence or a simple matter of low rentals if the Pakistani community grew from 600 to 10,000 inhabitants in the same years and the same areas where the huge Rambla del Raval has been built after the demolition of five blocks. They filled up the voided areas with their visible identity, and they found the opportunity to have a liveable public space, suited for a culture still used to spending an important part of the social life in streets or common spaces.[15]

In the centre of Barcelona there is a recognisable 'horizontal' subdivision of the urban tissue, where entire areas show clear signs of identity and are inhabited by specific groups of people. Maybe it is not accurate to talk about ghettoes, but the interchange among the different areas is not fluent, and in some cases difficult at least.[16] If this can be considered as the inevitable effect of an extreme transformation

---

14 This situation was actually forecasted in some of its consequences in López Sánchez, P., *El centro histórico: un lugar para el conflicto, estrategias del capital para la expulsión del proletario del centro de Barcelona. El caso de Santa Caterina y el Portal Nou* (Barcelona: Publicacions i Edicions Universitat, 1986).

15 Writer Manuel Vázquez Montalban described very well in his novels of the 1980s the atmosphere of old Raval, perceiving the transformations going on. More recently it has been possible to observe the effects of these changes in the theater. In 1999, local author Benet i Jornet described in the pièce 'Olors' the drama of a Barcelona family who must leave the territory where they have lived for generations, in 2002 a show called 'SuperRaval' represented a sort of Pakistan Superman solving the problems of the district. Recently, a movie, 'En construcción' (director José Guerin), and a music compilation, 'Barcelona Raval Sessions', highlighted the deep metamorphosis of the identity of this sector.

16 As we said, the surroundings of new Rambla del Raval have become a Pakistani sector with an important number of Maghreb people; in Raval North there's a huge community of Philippines; in the lower centre, South of Plaça Reial, are Moroccans and South Americans;

of the historical centre, another kind of subdivision is observable in Florence. Here we have a 'vertical' subdivision where there is not a clear territorial delimitation, or it is concentrated in few nodes. The point is that the same places are lived in, in very different ways by different groups of people at different hours and, thus, even the interpretation of the space changes considerably. Asian businessmen walking around early in the morning in Piazza Santa Maria Novella have nothing to do with the Somali women that, after working as domestic servants in private houses, meet together in the same square having social interaction and sometimes even the intimacy that is difficult to find in their small and too crowded homes.[17]

In the same way, rich Americans and Europeans who enjoy their coffee in Piazza Repubblica do not share the square in the same hours as the Albanians or Gypsy children. The weird effect is a kind-of urban schizophrenia that probably will last until the day the pressure of 'ideal renaissance town' identity, more affordable by rich immigrants, will force the poor ones and the original residents to move towards other sectors of the city.

The fact is that the discrepancy we are talking about is, apparently, solvable only in small and mid-sized cities, where the centre is not compressed between an ideal identity and the current one, considerably different and not always perceptible. Perhaps it is not really a matter of city size and it is depending on how many variables have been kept in mind by planners along the years.

It seems that the lack of a constant-planning-discussion-topic, under the point of view of the local identity (like the urgency to re-model in Barcelona or the Renaissance Golden Age in Florence) is the ideal condition in order to permit a flexible interpretation of the city. The famous Italian planner Bernardo Secchi says that an urban plan should be an open concept subdivided into many sub-areas improvable with projects on an architectonic scale.[18] Flexibility can give the opportunity for a maintenance of the historic centre characterised by a number of experiments: innovative solutions for car circulation; design of small interventions together with residents; a study of chromatic criteria for restorations; spurs for traditional stores; weekly markets brought once again to their original historical seats; efforts in merging the valorisation of the historic tissue and the main monuments together with contemporary works of art. All these actions can be efficient each time they are open to different interpretations. Not indifferent or neutral, then, but open.

Dominicans are concentrated around the old Santa Caterina market, close to the other Moroccan sector of Sant Pere. The Ribera is the seat of a composite population with Chinese, south Americans and Europeans from the EU. It is interesting to notice that the areas where the rehabilitation has been less destructive are the ones with a more proportioned presence among different communities.

17 See: Manlio Marchetta, 'La qualità dell'abitare e le funzioni urbane nell'area centrale di Firenze', La nuova città, VII 5/6 (1999): 135–140.

18 The work of professor Bernardo Secchi has often been focused on historic centres. In Italy, he planned the urban renewal for many big and small centres of ancient towns like Siena (1986–1990), Ascoli Piceno (1989–93) and lately Prato (1996–2000) described in detail in B. Secchi, *Un progetto per Prato, Il nuovo piano regolatore* (Firenze: Alinea, 1996).

Variations and variables in planning are not sufficient. The danger for the future is to create a 'ring-shaped cake' identity, where the natives all live around an abandoned centre, still full of symbolic meanings, even if void of life. And, as we have seen, it can be a void in both senses: because of an intense substitutive rehabilitation or for an obstinate and sterile preservation. Immigrants are, in this situation, only an occasional and temporary solution. They can fill up artificial urban voids and interstitial dismissed spaces or, if rich, they may transform the centres in an indistinguishable high-fashion mall. But when they build a community they develop their own way to read the town and its symbols, and they inevitably put new identities into historic cities.[19] The challenge lies in how the European historic centres will be able to recognise, accept and manage these new cultural values, that are different from the original ones but, above all, from a consolidated 'European' idea of modern urban development (a different use of public spaces and a different way of living at home). Somebody talks about urban 'ethnoscapes',[20] referring to the new inter-cultural city. Probably, in few years, the ethnoscapes of European historic centres will produce a new identity for those important symbolic spaces. Architects and planners can make a contribution in the designing of the multicultural town only if they have the skills to develop a flexible and open response to the problems of rehabilitation and valorisation of historic centres.

---

19 In the centre of Florence a lot of protestant churches are attended mostly by Asian immigrants, using these buildings as urban references. Barcelona Pakistanis have become the leading presence in Raval and the district is now immediately identified as their area. In many cities, Moroccans put their own names to the narrow central streets getting inspired by their morphologic features. A group of synchretists north-Brazilians prefer not to walk along certain ancient Florentine streets because, they say, these are 'full of dead' souls' they don't want to disturb.

20 'Ethnoscapes' is a term used for the first time in A. Appadurai, *Modernity at large, Cultural dimension of Globalisation* (Minneapolis: Univ. of Minnesota Press, 1996). Also see: Franco La Cecla, 'Metodologia della verità geografica', in C. Marcetti, N. Solimano and A. Tosi (eds), *Le culture dell'abitare* (Firenze: Edizioni Polistampa, 2003), pp. 17–28.

Chapter 10

New Urbanism and Planning History: Back to the Future

Christopher Silver

Christopher Silver is Professor and Head, Department of Urban and Regional Planning, University of Illinois, Urbana-Champaign (United States). His Ph.D. is in Urban History from the University of North Carolina, Chapel Hill. He was co-editor from 1993 to 1998 of the Journal of the American Planning Association and currently serves as editor of the Journal of Planning History. He is past-president of the Society for American City and Regional Planning History.

Introduction

At the annual conference of the Association of Collegiate Schools of Planning held in Atlanta in November 2000, the session which drew by far the largest attendance was 'Planning and the New Urbanism,' which featured architect Andreas Duany. Those in attendance were treated to an animated and, at times, vitriolic attack on planners and planning processes, not unlike the vinegary verbal assaults Jane Jacobs dished out in the early 1960s in *The Death and Life of Great American Cities*[1] and in her many reviews, essays and public presentations on planning. Like Jacobs, Duany decried the lack of consideration of human scale in the products of modernists, and challenged planners to revise the rules of development to create and to safeguard vibrant urban places.

What was distinctive about Duany's critique was that he celebrated a variety of past community designers and builders while bashing the ordinary planners as regulators. He contended in his Atlanta presentation, like he has in other writings, that the period prior to World War II (especially during the 1920s) was a golden era in community design and development. Planning was less institutionalized, subdivision regulations were based upon tradition rather than formulas, and community design schemes were the visionary products of landscape architects, architects and planning consultants given broad latitude by their private sector clients to shape urban places. New uniform community design standards perpetuated by the Federal Housing Administration beginning in the 1940s, coupled with the impact of the federal highway

---

1   Jane Jacobs, *The Death and Life of Great American Cities* (New York: Random House, 1962).

program on residential dispersal, created an automobile-dependent residential fabric devoid of interconnectedness and lacking civic virtue. Duany urged the planners and educators in the Atlanta audience to look backward for best practices and to revise the development rules for the future, to recapture lost elements from past planning as an antidote to sprawl.

What the New Urbanists offer, according to Duany, is a set of principles and range of development models to guide public policy, development practice, urban planning and design that counteract the pernicious effects of sprawl. The New Urbanist challenge is part of a broader critique of prevailing planning practices that seem to at the root of contemporary urban problems.[2] Like other environmentally-sensitive planners, New Urbanists regard the metropolis as the fundamental economic unit of the contemporary world, but see it is a finite place with clear edges that separate its inhabitants from farmlands, watersheds, open spaces and river basins. Development practices should sustain the metropolis as a whole through infill development and redevelopment that takes precedence over peripheral expansion. And these new developments should be communities that are physically compact, pedestrian-friendly and mixed use. Within these planned communities, a range of housing types and price levels can bring people of diverse ages, aesthetics and income into daily interaction, strengthening the person and civic bonds essential to an authentic community.

New Urbanism ideology calls for these compact communities to be linked through improved public transit and carefully designed streets that support interaction between civic and commercial centers and diverse residential areas. At the street level, the New Urbanism approach reinstates the local street as a human habitat rather than a machine mover or a barrier, with small streets providing controlled access within the neighborhood and with residential structured oriented to these narrow streets. The overall thrust of the New Urbanist approach is to resurrect a traditional neighborhood development modeled after the 1920s work and for the traditional neighborhood development to be the building block of a sustainable urban environment.

While some planners might find Duany's trenchant comments offensive (as they did with Jane Jacobs' critique in the 1960s), his message is music to the ears of the planning historian. Here is the principal of a leading planning and architecture firm with an impressive track record of successful community projects saying unbashedly that history matters, that planning history is important and that historical models can have an impact on the quality of contemporary life. Through the influence of Duany, 'New Urbanism has invigorated city planning history by invoking the tradition of American civic design to solve the conundrum of suburban sprawl',

---

2   James H. Kunstler, *The Geography of Nowhere: The Rise and Decline of America's Man-Made Landscape* (New York: Simon & Schuster, 1993); Jane Holtz Kay, *Asphalt Nation: How the Automobile Took Over America and How We Can Take It Back* (Berkeley: University of California Press, 1997); Roberta Brandes Gratz, with Norman Mintz, *Cities Back from the Edge: New Life for Downtown* (New York: Wiley, 1998); and Robert D. Bullard, Glenn S. Johnson, and Angel O. Torres, *Sprawl City: Race, Politics, and Planning in Atlanta* (Washington, DC: Island Press, 2000).

asserts planning historian Bruce Stephenson.[3] Of course, it is not all that unusual for architects to draw upon historical influences for the design of residential structures or whole communities for that matter. In many parts of the United States, history and tradition sell well. But what distinguishes the New Urbanist use of history is that it is employed not only to create a distinctive commodity but also to serve as a counterpoint to contemporary urban sprawl and a tool to recast prevailing urban development patterns. As Lee and Ahn assert in an assessment of Duany's celebrated Kentlands plan, the 'new urbanist ... paradigm challenges not only the prescriptive design standards and regulations governing suburban design but also the implicit values.' It 'represents a true value shift from contemporary suburban planning'.[4]

But this nexus between history and practice raises some interesting questions. What, for example, is the planning history that informs the New Urbanist critique and that constitutes the core of this paradigmatic shift? Virtually all of the leading publications of the new urbanists identify some specific historical influences but there are no two works that are entirely consistent in their acknowledgment of pivotal historical precedents.[5] One purpose of this study is to identify the explicit and implicit historical influences in New Urbanism. Not only is it unclear exactly which historical influences are most salient but also how this history is inbedded into New Urbanist principles and practices. Examination of a variety of New Urbanist authors makes this clear.

Another related and important question is how that history is used by New Urbanists. Is the New Urbanist application of planning history accurate, persuasive and interpretively sustainable? In this context, using planning and urban history inaccurately, without a sufficient rationale, and in such a way as to draw conclusions that cannot be sustained by the historical record might produce undesirable outcomes. The subtitle of this analysis, *Back to the Future*, is a reference to a series of movies by Steven Speilberg which explore the power of history. Marty McFly, the lead character played by Michael J. Fox, and his excessively frenetic sidekick, Doc Thompson, the inventor of a machine that propels them back and forth in time, both discover that it is not just knowing (or experiencing) history that determines the future. They were confronted with less savory unintended consequences that result from efforts to use the past in order to achieve other purposes. Simply recreating an imagined past by New Urbanists can also have unintended implications for current

---

3   Bruce Stephenson, 'The Roots of the New Urbanism: John Nolen's Garden City Ethic', *Journal of Planning History*, 1/2, (2002): pp. 99–123.

4   Chang-Moo Lee and Kun Hyuck Ahn, 'Kentlands Better than Radburn? The American Garden City and New Urbanist Paradigms', *Journal of the American Planning Association*, 69/1 (2003): pp. 50–71.

5   Peter Katz, *The New Urbanism: Toward an Architecture of Community* (New York: McGraw-Hill, 1991); Peter Calthorpe, *The Next American Metropolis* (New York: Princeton Architectural Press, 1993); William Fulton, *The New Urbanism: Hope or Hype for American Communities* (Cambridge, MA: Lincoln Institute of Land Policy, 1996); and Andres Duany, Elizabeth Plater-Zyberk and Jeff Speck, *Suburban Nation: The Rise of Sprawl and the Decline of the American Dream* (New York: Farrar, Straus and Giroux, 2000).

metropolitan development. The approach taken in this study is first to examine history as a component of the New Urbanist planning critique, then to see how planning history as precedent and precept is used in the creation of a New Urbanist community form, and, finally, to examine how planning history contributes useful lessons for New Urbanists.

History as Planning Critique

The New Urbanist use of history as a planning critique follows a relatively standard discourse of planning historians, a discourse which tends to focus on the limitations of professional applications. This is certainly evident in the magnum opus of New Urbanism, *Suburban Nation: The Rise of Sprawl and the Decline of the American* (2000), by Andres Duany, Elizabeth Plater-Zyberk and Jeff Speck.[6] The *Suburban Nation* is a saga about the victims of a half-century of bad planning, with particular emphasis on it suburban victims, including children ('the cul-de-sac kids'), alienated and detached women ('soccer moms'), and the 'stranded elderly', as well as their urban counterparts in 'bankrupt municipalities', namely the 'immobile poor'. The New Urbanists don't forget the 'hapless developer' who is required to operate against a stacked deck in their attempt to create liveable and profitable community in the face of insidious 'municipal regulations and engineering conventions'.[7]

The New Urbanist critique of planning's failures is consistent with other recent analyses of twentieth century urbanisation. As Robert Fishman observed in *The American Planning Tradition,* twentieth-century planning history is 'a record of missed opportunities and partial successes punctuated by outright failures.'[8](p. 5) At the same time, Fishman tenders notion of 'limited successes' as the misunderstood element of greatness in American planning. By this, Fishman implies that there has been rather remarkable changes introduced in urban areas through planning that have are a consequence that were less intentional but which have had enduring value nonetheless. Nonetheless, the New Urbanist critique is predominant view and has helped to identify those precedents that contributed to the current metropolitan malaise. These include the following:

1. City Beautiful Movement of the early twentieth century which gave birth to professional planning on the basis of which separating everything from everything else.[9]

---

6   Duany, Plater-Zyberk and Speck, *Suburban Nation: The Rise of Sprawl and the Decline of the American Dream.*

7   Ibid., p. 101.

8   Robert Fishman (ed.), *The American Planning Tradition: Culture and Policy* (Washington, DC: The Woodrow Wilson Center Press, 2000), p. 5.

9   Duany, Plater-Zyberk and Speck, *Suburban Nation: The Rise of Sprawl and the Decline of the American Dream*, p. 9.

2. The Federal Housing Administration (and the Veterans Administration) which financed large-scale housing development built on sprawl planning techniques.

3. The Federal Highway Aid Act of 1956 which routed highway 'directly through the centers of our cities, eviscerating entire neighborhoods and splitting downtowns into pieces' [10] and the automobile subsidy that diverted funds from transit.

4. Modernist architects designed towers in the park approach which relinquished the design of streets to engineers, and those that prepared the schematic for the modern suburb through proposals such as Radburn.

5. The new breed of bureaucratic planner after the 1930s that rejected history, aesthetics, and culture in favor of a process of planning and management relying exclusively on numbers.

The consequence of these dominant forces in American planning, as Duany, Plater-Zyberk and Speck put it, was that 'the American city was reduced into the simplest categories and qualities of sprawl.'[11] As Duany noted in his Atlanta presentation, a classic case of the output of contemporary American city planning is found in Virginia Beach, the largest city in Virginia in size and population. It is also the least densely populated place, a city of continuous sprawl with a center (not really a downtown in the classic sense) dominated by an eleven lane sea of asphalt supported by hundred of acres of parking. Other scholars have advanced similar critiques of contemporary urban development patterns, such as [12]Jane Holtz Kay, *Asphalt Nation: How the Automobile Took Over America and How We Can Take It Back* (1997), Richard Moe and Carter Wilkie, *Changing Places: Rebuilding Community in the Age of Sprawl* (1997) and William Kunstler, *The Geography of Nowhere* (1993) and *Home from Nowhere: Remaking our Everyday World for the 21st Century* (1996). So while New Urbanists are not alone in advancing such a harsh indictment of metropolitan development, their links to the development community make their views more likely to influence future practice than that of others.

History as Precedent and Precept

The historical antecedents that warrant careful consideration in understanding the New Urbanism precedents include a variety of traditional plans and enduring historic urban communities. Some pre-date the modern planning era, such as the

---

10 Ibid., p. 87.

11 Ibid., page 11.

12 Kay, Asphalt Nation: How the Automobile Took Over American and How We Can Take It Back; Richard Moe, and Carter Wilkie, Changing Places: Rebuilding Community in the Age of Sprawl (New York: Henry Holt and Co., 1997); James H. Kunstler, *The Geography of Nowhere*. (New York: Simon & Schuster, 1993); and James H. Kunstler, *Home from Nowhere: Remaking our Everyday World for the 21st Century* (New York: Simon and Schuster, 1996).

Figure 10.1 **One of the earliest built garden cities, Welwyn Garden City, provided a model for such influential American contributions to residential planning as the neighborhood unit plan of Clarence Perry**

Source: Stanley Buder, *Visionaries and Planners: The Garden City Movement and the Modern Community* (New York: Oxford University Press, 1990), Figure 14.

1794 plan by L'Enfant for Washington, DC, which was revived and implemented after 1900 and proved remarkably adaptable to the automobile age while maintaining low dependence on automotive infrastructure.[13] A prime example of the type of urbanisation advocated by New Urbanist but developed more incrementally rather than through a comprehensive plan is found in Alexandria, VA, a traditional walking city with the following key characteristics: a) a center; b) a five minute radius for each neighborhood cluster; c) a fine-grained street network utilizing a grid; d) narrow versatile streets; e) mixed uses; and f) special site for special buildings.

While evidencing a definite affinity for older cities which have preserved elements of their traditional neighborhood fabric, New Urbanism is appropriately connected to a rich tradition of suburban community planning that has its roots in Frederick Law Olmsted's plan and development of Riverside in the suburbs of Chicago in the 1870s. According to the Charter of the Congress of New Urbanism, the basic organizing elements of the metropolis are 'the neighborhood, the district and the corridor.' (Duany, Plater-Zyberk and Speck 1999, pp. 262–3).[14] One formative influence is Ebenezer Howard's Garden City since it offered a prototype of the balanced and self-sufficient community. New Urbanists distinguish the Garden City from the Garden Suburb, the latter being an incomplete version 'attached to an dependent upon a city'. But in fact there is an important connection to Garden Suburb that is more implicit than New Urbanists would want to admit, as I will discuss shortly.

Another important influence is the town planning schemes of early twentieth century planning consultant John Nolen, especially his plan for suburban Mariemont, Ohio, but also some of his community designs in Florida in the 1920s. James Kunstler made the Nolen connection in his 1993 book, *The Geography of Nowhere*,[15] noting that the plan for Seaside, which was the Duany/Plater-Zyberk signature New Urbanist project, 'was a modified neoclassical grid straight out of John Nolen' (p. 255). Duany has publicly acknowledged the Nolen influence but within the *Suburban Nation* there is little evidence that the authors consulted any of Nolen's writings or works by other scholars on Nolen. Research on Nolen's planning for small communities has effectively clarify the connection to New Urbanism but at the present it seems to be a more tenuous connection than some scholars would suggest.[16]

Less explicit but more important is the contributions of the regionalists of the 1920s (Lewis Mumford, Clarence Stein, Henry Wright and Benton MacKaye) with their notion of concentrated communities of a limited scale meshed together by transportation connectors. New Urbanist designer Peter Calthorpe (1993) identified

---

13 Duany, Plater-Zyberk and Speck, *Suburban Nation: The Rise of Sprawl and the Decline of the American Dream*, p. 7.

14 Ibid., pp. 262–3.

15 Kunstler, *The Geography of Nowhere*, p. 255.

16 Suzanne Sutro and Ronald Bednar, 'The Roots of Neotraditionism: The Planned Communities of John Nolen', *Proceedings Society of Society for American City and Regional Planning History*, (Richmond, Virginia: November 1991).

Ebenezer Howard's Garden City as a prototype of the transit oriented development that he advocates, namely 'towns ... formed around rail stations, with powerful civic spaces surrounded by village-scaled neighborhoods.'[17] This was contrasted by Calthorpe with the modernist vision of designer Tony Garnier, whose neighborhood model called for 'segregation of use, love of the auto, and dominance of private over public space' (p. 11).[18] Where the regionalists of the 1920s proclaimed the Garden city as the fundamental unit of the region, the New Urbanists stressed that 'the neighborhood is the elemental building block of the regional plan.'[19]

It is Clarence Perry's 'Neighborhood Unit' plan, one of the most influential outcomes from the 1927 Regional Plan for New York and Its Environs, that most fully captures the essence of the New Urbanist design scheme. Although not a planner or designer *per se*, but rather an educator by training, Perry was powerfully influenced by the Garden City planning model in devising his neighborhood unit plan. As historian Howard Gillette[20] pointed out in a 1983 essay, 'The Evolution of Neighborhood Planning', Perry knew what it meant to live in an American variation of Howard's Garden City since he moved his family into Forest Hill Gardens in 1912, a model community designed by Grosvenor Atterbury and the Olmsted Brothers under the sponsorship of the Sage Foundation Homes Company to introduce the Garden City to the American urban environment. Perry felt that Forest Hill Gardens contained all the 'essential elements' of the ideal neighborhood. It is important to note that Perry also was associated with the settlement house movement and in conjunction with the Russell Sage Foundation became an active participant in a national recreation movement aimed at improving park facilities to promote civic improvement. It was Perry's contention that expanding the functions of neighborhood schools to serve as neighborhood-based facilities for civic and recreational purposes was a way to strengthen the urban social fabric. In essence, Perry was advancing a notion not unlike the present social capital movement but without any direct link to local governance.

It was in a presentation to a joint meeting of the National Community Center Association and the American Sociological Association in 1923 that Perry unveiled the basic elements of the neighborhood unit plan. A more refined version of the neighborhood unit was published in the 1927 Regional Plan for New York. By the early 1930s, Perry's concept had been endorsed at President Herbert Hoover's National Conference on Home Building and Home Ownership as the best way to organise urban areas.[21] By the 1940s, through the advocacy of national planning consultant Harland Bartholomew, the neighborhood unit plan had been fashioned

---

17 Calthorpe, *The Next American Metropolis*, p. 11.

18 Ibid., p. 11.

19 Duany Plater-Zyberk and Company, *The Lexicon of New Urbanism* (Miami: Duany Plater-Zyberk and Company, 1997).

20 Howard Gillette, 'The Evolution of Neighborhood Planning: From the Progressive Era to the 1949 Housing Act', *Journal of Urban History* 9/August (1983): 421–444.

21 Ibid. pp. 422–427.

Figure 10.2 **Clarence Perry's neighborhood unit plan, most fully expressed in volume seven of *The Regional Survey of New York and Its Environs* (1929) was a formative influence on urban residential development and the conceptual basis for public housing communities as well as private developments.**

Source: Clarence Perry, *Neighborhood and Community Planning* (New York: *Regional Survey of New York and Its Environs*, 1929).

into a comprehensive urban revitalisation strategy anD provided the scale model for neighborhood revitalisation through the urban renewal program in the 1949 Housing Act.[22]

The organisation and spatial arrangement of Perry's neighborhood unit is directly replicated in the new urbanism planning unit. In the neighborhood unit and the New Urbanism traditional neighborhood development, size is 'determined by the walking distance of five minutes from center to edge'. The neighborhood unit plan utilised a civic center with the elementary school as its anchor. A pattern of small streets within the neighborhood dispersed local traffic and the placement of retail and regional institutions at the edge to support the use of automobiles (without requiring automobile traffic to penetrate the neighborhood) were key elements of Perry's

---

22 Christopher Silver, 'Neighborhood Planning in Historical Perspective', *Journal of the American Planning Association*, 50/Spring (1984): pp. 161–174, pp. 168–169.

neighborhood unit plan. All of these were elements picked up by New Urbanists in their traditional neighborhood development. The differences are negligible. For example, the New Urbanists shifted the school to neighborhood edge, expanded space for regional institutions at the edge, and in the center replaced the neighborhood school and the open space playgrounds with a mixed-use civic-commercial center. This mixed use civic-commercial center linked to the edge through the street pattern and was supported by abundant parking.

Perry's abiding concern was to adapt city neighborhoods to the automobile age. He was convinced that 'the automobile has been ... a destroyer of neighborhood life' and that a growing number of expressways were 'cutting up residential areas into small islands separated from each other by raging streams of traffic', and so the design of the neighborhood unit plan ensured that the automobile remained largely at the neighborhood edge. While Perry's neighborhood unit sought to exclude the automobile, the automobile, in effect, supplied the long-awaited rationale for neighborhood-focused planning that he advocated. The New Urbanist traditional neighborhood development inherited the automobile-induced sprawl (and transit-less cities) and like the neighborhood unit plan used this as a rationale for a new

Figure 10.3   **Apart from the initial project of Seaside, Kentlands, Maryland was the most ambitious New Urbanism community designed by Duany and Plater-Zyberk, and one that was situated in a dynamic metropolitan region as distinguished from the resort community of Seaside.**

model. Yet while fundamentally opposed to automobile sprawl, as previously noted virtually all New Urbanist projects have been designed around the automobile. Even Calthorpe's transit-oriented neighborhood model did not preclude reliance upon automobiles, and could function just as well without mass transit. From Seaside to Celebration, Florida, and Kentland, Maryland to Laguna West, California, the New Urbanist operated predominantly in an auto-center, upscale suburban environment.[23] While these important projects advanced new models of neighborhood development, they have largely been confined to expanding the options available to suburbanites rather than fostering the diversity and sustainability of the compact metropolis.

Critical Lessons from History

Despite the central place of Perry's neighborhood unit plan in the New Urbanist lexicon, it is remarkable that the core literature of Perry and his proponents[24] as well as the critiques that dogged the neighborhood unit plan throughout the post-World War II era, are missing from the New Urbanist writings. What is also missing is a critical assessment of the social implications of the neighborhood unit plan. Perry's social agenda was not only aimed at protecting residents from the deleterious effects of the automobile but also from the social diversity inherent in dynamic urban environments. Through the federal Housing and Home Finance Agency, Perry's neighborhood unit became a neighborhood prototype that was 'an instrument for segregation of ethnic and economic groups', according to Chicago planner and activist, Reginald Isaacs writing in 1948.[25] Isaacs astutely recognised, as Perry himself had stated in earlier writings, the self-contained neighborhood was desirable only so long as it protected 'widely different classes and races' from living side by side. Perry's neighborhood unit would safeguard social homogeneity in a number of ways, by preventing the increase of classes that are beyond the reach of housing reform, by providing a physical barrier against the encroachment of deteriorated neighborhoods, and by providing an object lesson in improved housing that could be modeled throughout the metropolis. Critics of New Urbanist development point to the same tendencies to construct communities that are socially, economically and racially monolithic, and where the goal is not necessarily to integrate with the existing urban fabric but to remain separated and distinctive.

Besides failing to recognise how the neighborhood unit plan reinforced social and economic segregation in the American city, New Urbanism falls into the trap that so effectively negated the virtues of Radburn and Perry's neighborhood unit plan, namely concentrating on design as an end in itself to create livable and

---

23 Fulton, *The New Urbanism: Hope or Hype for American Communities*.

24 Clarence Perry, *Housing for the Machine Age* (New York: Russell Sage Foundation, 1939); and J. Dahir, *The Neighborhood Unit Plan* (New York: Russell Sage Foundation, 1947).

25 Reginald Isaacs, 'The Neighborhood Unit as an Instrument for Segregation', *Journal of Housing*, 5/August (1948): pp. 215–9.

cohesive communities. The environmental determinism of the New Urbanism is a direct descendant of the neighborhood unit plan proponents, although it is tempered somewhat by statements that proclaim more inclusive social and economic objectives. But those places that have been the recipients of New Urbanism approaches are virtually all in greenfield locations in middle and upper income markets where the likelihood of achieving inclusionary objectives is minimal. The history of Garden cities, new towns, new communities, urban renewal and suburban building in American cities over the course of the twentieth century should have pointed out to New Urbanists how new self-contained community developments often fall far short of intended social and economic objectives. The histories of suburbanisation and new community development provide ample evidence to understand the inherent pitfall of this approach to urban development. In addition, the histories of participatory community development, neighborhood-based planning and an array of housing and social reform movements offer a note of realism that is lacking in the New Urbanist formula. Unfortunately, there appears to be too little attention among New Urbanists to the abundant planning history that demonstrates the fragility of similar enterprises in the past.

Urban History as Planning Precedent

There is an implicit assumption in the New Urbanist use of historical antecedents that these largely involve comprehensive models of urban development, such as those of Perry, Nolen, Howard and Olmsted. In fact, these comprehensive planning models account for a relatively small proportion of the community development efforts of the nineteenth and early twentieth centuries. The work of urban historians, which is largely absent in the New Urbanist literature, offers a valuable counterpoint by demonstrating that urbanisation in the United States was largely a result of an incremental process of urban accretion. The classic study of the origins of suburban Boston by historian Sam Bass Warner, *Streetcar Suburbs: The Process of Growth in Boston, 1870-1900* (1962),[26] as well as a more recent study of Chicago by Ann Durkin Keating, *Building Chicago: Suburban Developers and the Creation of a Divided Metropolis* (1988)[27] demonstrate through detailed analysis of places that new urban neighborhoods were fashioned by hundreds of small scale builders taking advantage of the transportation innovations of the late nineteenth century to situate and market their products. They functioned as urban planners in the absence of more formal institutions, and differed fundamentally in motives and methods from those involved with more comprehensive approaches. What is important is that their communities relied **exclusively** (emphasis added) on a mass transit system for linkages to the larger urban region, with local interactions confined largely to

---

26 Sam Bass Warner, Jr., *Streetcar Suburbs: The Growth of Boston, 1870–1900* (Cambridge: Harvard University Press, 1962).

27 Ann Durkin Keating, *Building Chicago: Suburban Developers and the Creation of a Divided Metropolis* (Columbus: Ohio State University, 1988).

Streetcar Buildouts   o)

HANDSOME BUSINESS AND RESIDENCE LOTS.

On Grand Boulevard, a beautiful 80 foot macadamized and sewered street extending through the property and lighted for its entire length by street lamps.

ONLY 100 LOTS LEFT FOR SALE IN THESE BLOCKS.

South-east corner Dearborn and Randolph Sts.

Figure 10.4 **This plan of Grossdale, a streetcar suburb in Chicago in the 1890s, demonstrates the early links between mass transit and suburbanisation and the high density of early residential suburban models.**

Source: Dolores Hayden, *Building Suburbia: Green Fields and Urban Growth*, 1820–2000 (New York: Pantheon Books, 2003), p. 85.

the traditional walking city model. Since so many of these neighborhoods were fashioned prior to the widespread application of land use regulations in cities, and zoning in particular, it was possible and highly desirable to intermingle residential and non-residential uses, rather than to impose the rigid segregation of land use that later accompanied the proliferation of zoning. Absent the restrictions imposed by zoning and later subdivision regulations, it was tradition and function that dictated how this mixed land use pattern emerged.

Where the New Urbanists have uncovered the real problem of the perpetuating urban dysfunction is the local development rules and planning codes that reinforce the larger forces leading to sprawl. Duany's stinging critique of the rules by which local planners regulate development is supported by much of the work done by urban and suburban historians in dissecting the urban development processes of nineteenth and twentieth centuries.[28] And yet this critical aspect of New Urbanism has been greatly overshadowed by the focus on designing comprehensively planned communities amidst the current suburbanisation processes. Dolores Hayden's recent historical analysis of suburban development, *Building Suburbia: Green Fields and Urban Growth, 1820-2000* (2003)[29] suggests that 'preserving historic layers in suburbia,' namely the suburbia created from the late nineteenth century through the 1920s, must be the key focus of a 'new urbanism,' not new development however stylish in the greenfields. As she notes,

> hundreds of books and articles discuss the tools for 'smart growth' available to suburbs. They all advocate regional land use planning, but most of them do not include much analysis of how suburbs have developed over time. Historical analysis of different layers of suburban neighborhood development can contribute to preservation by defining the distinctive quality of different places.[30]

And this of course requires interventions that are sensitive to context of when those places were developed. 'Knowledge of how their historic cultural landscapes have evolved can often help to establish priorities for current interventions,' Hayden concludes.[31]

A New Urbanist Legacy

The real challenge for New Urbanism is not just to redesign the new suburban model but to tackle of the problems of the old urbanism, and to reintroduce viable community

---

28 Sam Bass Warner, Jr., *The Urban Wilderness: A History of the American City* (New York: Harper and Row, 1972); and Eric H. Monkkonen, *America Becomes Urban: The Development of U.S. Cities and Towns, 1790-1980* (Berkeley: University of California Press, 1988).

29 Hayden, Dolores, *Building Suburbia: Green Fields and Urban Growth, 1820-2000* (New York: Pantheon Books, 2003).

30 Ibid., p. 235.

31 Ibid., p. 235.

forms in existing urban areas. In its current mode, New Urbanism reinforces the basic components of sprawl. William Fulton notes that some New Urbanists have recognised that there needs to be more to the movement than designing better suburban neighborhoods. This wing of the New Urbanist movement points to key inner city redevelopment projects, such as Battery Park City in New York, Ghent Square in Norfolk, Virginia and Seattle Commons in Seattle, Washington as models of desirable infill developments, as well as the successful projects to revamp public housing along New Urbanist lines, such as in Digg Town in Norfolk. Accordingly, New Urbanism principles can and should be applied across a broad spectrum of situations not just emerging suburbs, but also in underutilised and abandoned industrial sites, struggling inner-ring suburbs, and small towns outside the metropolis altogether.

Examining the narrow, somewhat misused, and neglected historical legacy of cities may help us to understand why New Urbanism in its most prevalent manifestations has not succeeded in combating sprawl, or in creating the broader vision of the humane metropolis that it espouses. Building a 'new urbanism' on a much broader base of historical inquiry is the essential message of this study. It may lead to a new planning as well as a new urbanism. There are histories that can make a difference, and it is hopeful as we create new urban futures that we draw upon a broader and less romanticised version of that planning past as our guide.

PART III
Cultural Urbanism and Planning Strategies

Chapter 11

Branding the City of Culture – The Death of City Planning?

Graeme Evans

Graeme Evans is Professor of Urban Cultures and Director of the Cities Institute at London Metropolitan University (England). His research covers cities and cosmopolitan cultures, urban design and planning, cultural policy and the creative industries. He is leading a comparative study of 'Creative Places' with case studies in Barcelona, Berlin, New York and Toronto, and is author of 'Cultural Planning: An Urban Renaissance?'

Introduction

The planned city – if such a technocratic notion exists today outside of Fritz Lang's mythical 'planners who dream whilst others toiled' (*Metropolis 1926*) – has increasingly given way to the equally self-defeating branded city. This is reflected in old world cities and post-industrial cities of manufacturing and docklands renewal, to new world and emerging cities that seek to join the global network, not just through rapid urbanisation, but through cityscapes that have more resemblance to a film set which masks their underbelly and poverty, from Kuala Lumpur to Shanghai.[1] Whilst urban place-making and city marketing has been practiced by public and private boosterists since the nineteenth century,[2] and by empire builders for thousands of years before, the explicit branding of whole cities in a similar manner to corporate global brands is now apparent[3] as cities strive for a distinctive tag and image that can satisfy the footloose tourist, investor, and members of the creative class alike.

---

1   R.K. Biswas (ed.), *Metropolis Now* (Vienna: Springer-Verlag, 2000); F. Wu, 'The global and local dimensions of place-making: remaking Shanghai as a world city', *Urban Studies*, 37 (2000): pp. 1359–1377; and F. Gilmore, 'Shanghai: Unleashing creative potential', *Journal of Brand Management*, 11/6 (2004): pp. 442–448.

2   S. Ward, *Selling Places: The Marketing and Promotion of Towns and Cities 1850–2000* (London: E & FN Spon, 1998); and G. Ashworth and H. Voogd, H. 'Marketing and Place Promotion', in J.R. Gold and S.V. Ward (eds.), *Place Promotion: The Use of Publicity and Marketing to Sell Towns and Regions* (Chichester: Wiley, 1994).

3   M. Kavaratzis, 'From City Marketing to City Branding', *Journal of Place Branding*, 1/1 (2004): pp. 58–73.

Indeed, Florida's 'creative class'[4] are now presented as the 'storm-troopers'[5] of city branding.

The development of city-centre and redundant industrial and docklands areas of cities, creating new settlements, festival waterfronts, and centres of government control and business, can be seen as an early form of branding the city by the imposition of the post-medieval grid, boardwalk and promenade. In the past these reclaimed city centre spaces housed national institutions – government offices, palaces, museums and galleries and public squares, such as the imperial palaces in Vienna and Paris, the museum island in Berlin. They also created a grid 'brand' which was exported to the colonised Americas, as well as in new towns and cities irrespective of their topography or need for the militaristic control which drove Haussman's masterplan for Paris, and Cerda's scientific rationality applied to Barcelona's 1850s city plan.[6] Nineteenth century colonial occupiers thereby 'consciously manipulated the urban landscape to symbolise and reinforce their claims to legitimate rule'.[7] Examples are evident in capital and regional cities of Latin America, combining the characteristics of pre-colonial cities, which have been adapted to modernisation in Bogata and Sao Paolo. Most Spanish cities in the 'New World' were laid out in terms of the Laws of the Indies (1573), specifying the grid-iron street plan already implemented in cities such as Puebla in the 1530s, with central plaza and church/cathedral and walled building plots, replicated on a smaller scale throughout the city, and facilitating religious and state control.[8]

Corporate City Branding

What these clearances also generated was huge commercial gains for developers and landlords, which are reflected today in the beaux quarters and high value property in uptown business, apartment, hotel and retail districts. This pattern is evident in regeneration sites in London's Docklands, La Défense, Paris and New York's Battery Park, and in gentrified areas of the former working class districts of Barcelona, Bilbao and Berlin. What this branding of commodified space also shares over time is the fact that little or none of the increased values accrue to the residents who have

4   R. Florida, *Cities and the Creative Class* (New York: Routledge, 2005).

5   'Artists are the storm-troopers of gentrification' a Montreal graffiti-artist's quotation reproduced on the cover of the Toronto Arts Council's annual report (TAC 1988), referring to the subsequent development of artists studios and lofts for commercial property development and gain.

6   M. Miles, 'Drawn and Quartered: El Raval and the Haussmannization of Barcelona', in M. Jayne and T. Bell (eds.), *City of Quarters: Urban Villages in the Contemporary City* (Aldershot: Ashgate,2004), pp.37–55.

7   T. G. Jordan-Bychov and M. Domosh, *The Human Mosaic: A Thematic Introduction to Cultural Geography* (New York: Addison-Wesley, 1999), p.412.

8   G.L. Evans, *Cultural Planning: An Urban Renaissance?*, (London: Routledge, 2001).

Figure 11.1
La Défense and Louvre

been decanted and displaced, i.e. a 'Pareto loss' and a failure to distribute 'planning gain'. Today museum quarters, heritage districts such as Venice and Florence, and modern *grands projets*, reinforce the city brand whilst limiting access and development. This includes the use of conservation area/building planning powers to resist contemporary development and alternative identities, notably non-European heritage.[9] The development of La Défense and the Grand Arche on the axis to the Louvre for example, emphasises the city grid, whilst I.M. Pei's Louvre extension houses an underground shopping plaza-entrance, *Carousel du Louvre*, drawing on one of the first leisure-shopping experiences under one roof – the Bon Marché.

This 'symbolic externality' manifested through property and urban design 'hallmarks' increasingly displaces notions of the vernacular, of a sense of place, and prosaic amenity planning for residents and commuting workers, for whom the city 'brand' is at best, a tired cliché (e.g. Macintosh, Glasgow; Gaudi, Barcelona); a misrepresentation by public officials/politicians; or at worst a wholesale clearance of urban areas. This occurs through large scale redevelopment incorporating culture-led regeneration[10] through the crowding-out of incumbent communities, as infrastructure, housing and employment is subjected to extreme gentrification effects. For example El Born and Raval districts in Barcelona; La Vieja, Bilbao; Kreuzberg, Berlin; Shoreditch and Spitalfields, east London. Mommaas therefore identifies three conflicts arising from the practice of city branding.[11]

1. the tendency to gear city brands to the dynamic of an external cash-rich market rather than to that of internal cultural practices and feelings;
2. the tendency to objectify and generalise specific cultural meanings by means of 'brands' and then to link these meanings materially to spectacular places and projects;
3. the possible danger that 'brands' preclude renewal rather than stimulate it.

In contemporary city branding strategies and projects, all three of these scenarios are evident, not least in the areas of cultural consumption and iconic developments which emulate product branding strategies. This particular manifestation of globalisation is on the one hand the physical representation of product promotion and placement already transmitted through advertising, sponsorship and broadcast and digital media, as well as a rational economic strategy to 'cut out the middle men' of general retailers and reduce distribution and transaction costs. It also presents an opportunity for enhancing customer loyalty and synergy between the physical

---

9   G.L. Evans, and J. Foord, 'Rich mix cities: from multicultural experience to cosmopolitan engagement', *Ethnologia Europaea: Journal of European Ethnology* (2005) (forthcoming).

10 G.L. Evans, 'Measure for Measure: Evaluating the Evidence of Culture's Contribution to Regeneration', *Urban Studies*, 42/5–6 (2005): pp. 959–983.

11 H. Mommaas, 'City Branding', in T. Hauben, M. Vermeulen and V. Patteeuw, *City Branding: Image Building and Building Images* (Rotterdam: NAi Publishers, 2002), p. 38.

and symbolic value of brands. In this respect the branded city also bypasses the nation-state and associated national identity, not least in the cosmopolitan city with international (global) aspirations.[12]

It is the twin value systems of the political and symbolic economy that according to Zukin[13] provides the most productive analyses of the built environment of cities. However the branded entertainment-retail emporium and art museum complex challenges normal planning-use, design and vernacular preferences – is it a shop, a venue, an arcade, advertisement, public art or monument? The contemporary EXPO site, a long established branding event along with Olympic and other major sporting events, falls into this ambiguous 'non-place' – for example Lisbon's leftover 1998 EXPO, London's Millennium Dome at Greenwich, Montreal's 1967 EXPO and Barcelona's 2004 UNESCO Forum complexes.[14,15] In securing international designation and imported brand status, the actual places they occupy and their disfunctional and transient nature, presents one paradox. Another is the competitive drive for such brands and branded events (despite this past experience) which cities continue to display, in contrast to the traditional notion of the planned city and the mediation of development. Their prominence and scale thus accelerates the hard branding of cities already struggling to resist a(nother) McDonald's, Wal-Mart, mall or multiplex, and now Guggenheim franchise (below), which are generally developed at the cost of existing, independent or at least more widely distributed outlets and town centres such as cinemas, museums, speciality retail stores, parks and squares.[16]

City Branding – Past and Present

Towns and cities have of course long been identified with major corporate headquarters, factories, or sporting venues and clubs, and particular producer associations, such as Silicon Valley, Motown, Nashville, Hollywood, 'Guinness World' Dublin, much as rural towns associate themselves with their agricultural produce and annual harvest festivals, but these are more marketing and civic imaging devices than wholesale city

---

12 Evans, *Cultural Planning: An Urban Renaissance?*

13 S. Zukin, 'Space and symbols in an age of decline', in A. D. King (ed.), *Re-Presenting the City: Ethnicity, Capital and Culture in the 21st Century Metropolis* (London: Macmillan, 1996), pp. 43–59, p. 43.

14 G.L. Evans, 'Hard Branding the Culture City – From Prado to Prada', *International Journal of Urban and Regional Research*, 27/2 (2003): pp. 417–440.

15 Evans, 'Measure for Measure: Evaluating the Evidence of Culture's Contribution to Regeneration'.

16 G.L. Evans, 'Urban Leisure: Edge City and the New Pleasure Periphery', in M. Collins and I. Cooper (eds.), *Leisure Management: Issues and Applications* (Wallingford: CAB International, 1998), pp. 113–138.

repositioning and place-making.[17] Culture cities also seek to develop this association amongst visitor and resident, since as Mommaas observes:

> Brands derive their attraction largely from the fact that they introduce a certain order or coherence to the multiform reality around us. Brands enable us more easily 'read' each other and our environment of places and products. Seen in this way brands are not purely a source of differentiation, but also of identification, recognition, continuity and collectivity.[18]

As Simmel suggested, branding city quarters in the past provided a link between the diverging individual and collective culture and identity, reconnecting the locale with a sense of socio-cultural 'belonging', whether to a city, neighbourhood or nation.[19] Increased mobility, wanderlust and search for new products and experiences were already apparent in the nineteenth century, and as Zukin maintains today:

> the cultural power to create an image, to frame a vision, of the city has become more important as ... traditional institutions have become less relevant mechanisms of expressing identity. Those who create images stamp a collective identity. Whether they are media corporations like Disney, art museums, or politicians, they are creating new spaces for public cultures.[20]

Simmel followed in the visual 'flaneur' tradition of Baudelaire and Benjamin, but it is perhaps the production and usage of everyday space which Lefebvre distinguished,[21] and which is absent from these experiential perspectives of the city. Branding seeks to create a unified version, to make concrete the symbolic and habitus, but unlike the branded farm animal, ownership cannot be reduced in this way. Nor can it replace spatial planning and more democratic city management. City ownership and identities are complex and subjective, particularly in relation to national and ethnic roots, but also in terms of neighbourhood, district and group identification.

One brand-inspired response to this, is cities which seek to be both inclusive and project their multicultural 'rich mix'.[22] This is done by 'rediscovering' and re-labelling their ethnic quarters which had come to be associated with deprivation and decline – Jewish ghettos and immigrant neighbourhoods such as Greektown and Little Germany – and which are now branded through cultural association, despite the fact that their ethnic communities have long gone to the suburbs (e.g. Little Italy, Toronto). Examples of these ethnoscapes[23] include *Banglatown*, East London;

---

17   S. Ward, *Selling Places: The Marketing and Promotion of Towns and Cities 1850–2000*

18   Mommaas, 'City Branding', p.34

19   Ibid.

20   S. Zukin, *The Cultures of Cities* (Cambridge. MA.: Blackwell, 1995), p. 3

21   H. Lefebvre, *The Production of Space* (trans. Nicholson-Smith, D.) (Oxford: Blackwell, 1974); and H. Lefebvre, *Critique of Everyday Life* (trans. Moore, J.) (London: Verso, 1991).

22   K. Worpole and L. Greenhalgh, *The Richness of Cities: Urban Policy in a New Landscape – Final Report* (Stroud: Comedia/Demos, 1999).

23   A. Appadurai, *Modernity at Large: Cultural Aspects of Globalization* (Minneapolis: University of Minnesota Press, 1997).

Curry Triangle, Birmingham and Bradford; and the ubiquitous Chinatowns. The latter are both recreated and multiplied (e.g. in Toronto and Manchester) or even used as a generic tag to indicate a bohemian quarter prior to regeneration such as the *Barrio Chinois* in El Raval district of Barcelona, which never actually hosted a Chinese community – more the oriental 'other'. The cosmopolitan city brand is also a competitive advantage which world cities such as London, New York and Toronto boast in terms of the numbers of nationalities represented and languages spoken. Toronto's new city marketing brand 'City Unlimited' in part reflects this diversity, an identical strapline which London had at the same time rejected (being 'unlimited' also exposes a corporation to a lack of protection from liabilities, i.e. debts, arising from insolvency). In Singapore 'Global City of the Arts' on the other hand, its Anglo-Chinese bilingualism underscores its position between the great trade opportunity of mainland China and the West.[24] The melting pot metaphor associated with cosmopolitan cities such as London and New York, is used both to promote itself and to capitalise on the competitiveness of such diversity in global markets and communications. In London's case this linguistic and cultural diversity was used effectively in the final presentation for the city's 2012 Olympic bid to the IOC in Singapore. A day later, terrorist bombs took over 50 lives, many of them representatives of these ethnic communities, including students, tourists and British born muslims. Dividing these communities may have been one goal of the terrorist, whose own brand of fundamentalist monoculturalism also rejects the dominant late-capitalist ideology and way of life.

Over-reliance on a single brand can also risk image decay as the brand dilutes, so as Bilbao's Provincial president, Josu Bergara says, seemingly with no hint of irony: 'Other cities will have to find their own projects, not copies of the Guggenheim'.[25] As Giddens remarked: 'Money and originality of design are not enough. You need many ingredients for big, emblematic projects to work, and one of the keys is the active support of local communities'.[26] According to the Guggenheim Foundation: 'the [Guggenheim] sites are distinct and distinctive entities. Each is a unique exhibition space that responds to and is shaped by its particular environment and constituency' (website). However, shortly before the competition for the design of the Guggenheim museum was awarded by the regional government, they had turned down the funding for a Basque Cultural Centre in the centre of the city, and subsequently reduced its subsidies to a broad range of cultural groups in the region.[27] One objective that

---

24   T.C. Chang, 'Renaissance revisited: Singapore as a 'Global City for the Arts'', *International Journal of Urban and Regional Research*, 24/4 (2000): pp. 818–831.

25   L. Crawford, 'Bilbao thrives from the 'Guggenheim effect', *Financial Times Weekend*, London, April 28, 2002: p. 2.

26   Crawford, 'Bilbao thrives from the 'Guggenheim effect'', p. 2.

27   M.A. Gomez and S. Gonzalez, 'A Reply to Beatriz Plaza's 'The Guggenheim-Bilbao Museum Effect'', *International Journal of Urban and Regional Research*, 25/4 (2001): pp. 888–900; and E. Baniotopoulou, 'Art for Whose Sake? Modern Art Museums and their role in Transforming Societies: the Case of the Guggenheim Bilbao', *Journal of Conservation and Museum Studies*, 7 (2000): pp. 1–15.

Figure 11.2 Guggenheim Bilbao

Guggenheim has yet to achieve for Bilbao is the return of Picasso's *Guernica* to the Basque country, perhaps the only authentic gesture which would link this flagship to any notion of regional identity – only 7 out of 53 acquisitions in the first three years were from Basque artists. Guggenheim Bilbao is now experiencing a decline in visitor numbers, with low-cost flights being cut back. Its dependence on tourists (few local visitors go there) will accelerate a crisis in this particular franchised brand: 'Bilbao looks great now, but in 30 years will it prove less flexible?'[28]

Grands Projets

What many of these iconic interventions also have in common, is an exorbitant cost, often turning out to be overdue and overbudget, some overrunning by over 100%, and a defiance of public planning and choice: 'which large projects get built? ... those for which proponents best succeed in designing-deliberately or not – a fantasy world of underestimated costs, overestimated revenues, overvalued local development effects, and underestimated environmental impacts'.[29] Like major hallmark event projects, iconic culture-led regeneration schemes are not wholly grounded or rational decision-based. They rely more on blind faith, 'pork barrel politics'[30] and constructed visions which appear not to look beyond the short-term physical impacts and landscapes they create.[31] Cities such as Rio are consequently resisting the Guggenheim franchise, estimated to cost the city $250m (judges have declared the building contract between mayor and Guggenheim null and void), a similar position taken from new regeneration areas such as the Thames Gateway in South East England:

> The landmark building has become a staple element of urban regeneration, especially in industrial locations. But not every town can sustain its own Tate Modern, and the long term sustainability of such iconic statements is being increasingly questioned...the iconic building as regenerative catalyst may be the wrong answer.[32]

For instance, over £3 billion in capital investment has been made in Paris with more than 600 provincial *Grands Travaux* projects costing in excess of a £250 million. Despite their financial burden and uncertain impact, Mitterand used these *Grands Projets* to 'rally behind a certain idea of the city'.[33] In particular to reconcile what he saw as the divorce of architecture from town planning which had given rise to two undesirable urban phenomenon: tower blocks and as a reaction to this, 'dispersion'. According to Scalbert however, these projects have actually reinforced the centre-periphery divide: 'whatever their value as architectural set-pieces, they are not the much-vaunted harbingers of a proclaimed urban renaissance. On the

---

28   M. Irving, 'Museum trouble', *Blueprint*, 156 (1998): pp. 26–8, p. 28.

29   B. Flyvbjerg, 'Designing by Deception: The Politics of Megaproject Approval', *Harvard Design Magazine*, 22 (2005): pp. 50–9, p. 50.

30   D. Sudjic, *The 100 Mile City* (London: Flamingo, 1993), p. 31.

31   Evans, 'Measure for Measure: Evaluating the Evidence of Culture's Contribution to Regeneration'.

32   Charrette 3, *New Models of Cultural Facilities* (Tilbury, Thurrock: 26 May, 2004), p. 3.

33   F. Mitterrand, 'Preface', in E. Biasni, (ed.), *Grands Travaux* (Paris: Connaissance des Arts, 1989), p. 5.

contrary, like circus games, they direct attention from the inexorable erosion of Paris and the brutal neglect of its suburbs'.[34] The housing estates of outer Paris had been the source of the term 'social exclusion', which has found its way into major policy rhetoric and programmes in the UK and North America (*New Deal* etc.). With a 'Non' vote to the new European Constitution, and unexpectedly losing to London in the 2012 Olympic bid, how Paris and other French cities will respond remains to be seen – more *Grands Travaux*, more rebranding? City planning in this context seems too benign and long term a proposition for political and commercial market timespans.

During the 1980s/1990s the use of culture as a conduit for the branding of the 'European Project' has added fuel to this culture city competition, whilst at the same time celebrating an official version of the European urban renaissance. The brand in this case is manifested by the EU logo and packaged as the European culture city-break and contemporary Grand Tour,[35] in a similar manner to UNESCO's branded world heritage sites. However, as the recent ECC assessment found: 'too often, Capitals of Culture have focussed most of their efforts on funding of events and projects that form part of a year-long celebration, with too little time and investment given to the future'.[36] Although the ECC programme was a self-conscious extension of the 'European project', the first *City of Culture of the Americas* was held in Merida, southern Mexico in 2000, whilst international creative city league tables are now rationalised by boho (bohemian), innovation and technology, 'openness' and other quality of life indices.[37] However, in some respects these essentially lifestyle rankings which correlate selected location advantages with creative milieus, are also associated with socio-cultural and spatial inequalities and gentrification effects.[38]

In the case of prospective Olympic bids, cities such as Manchester and Toronto have staked the regeneration of major sites (e.g. Toronto's waterfront) on unsuccessful failed bids, often despite popular resistance, whilst others have been burdened by the financing and failure associated with undeveloped post-event sites, in some cases long after they occurred (e.g. Montreal 1967 EXPO and 1976 Olympics; Athens 2004 Olympics). The reality is that even in 'successful' (sic) games such as Barcelona and Sydney, post-event facility usage has been unsustainable and even a burden on local residents who inherited such unplanned facilities. Sydney's new Olympic surburbs

---

34   R. Scalbert, 'Have the Grands Projets really benefited Paris?', *Architect's Journal*, 3/200 (1994): p. 20.

35   G.L. Evans, 'In search of the cultural tourist and the postmodern Grand Tour' International Sociological Association XIV Congress, *Relocating Sociology* (Montreal, July 1998).

36   R. Palmer, *European Capitals/Cities of Culture. Study on the European Cities and Capitals of Culture and the European Cultural Months (1995–2004)* Part I and II, Palmer/Rae Associates. (Brussels: European Commission, 2004), p. 5.

37   Florida, *Cities and the Creative Class*; and T. Nichols Clark, *The City As Entertainment Machine* (Oxford: Elsevier, 2004).

38   Evans, 'Measure for Measure: Evaluating the Evidence of Culture's Contribution to Regeneration'.

have failed to deliver the sustained community they planned – residents have to drive to the city for work. In London2012's successful Olympic bid, the regional masterplan for the swathe of east London along the (river) Lee Valley plans new housing and amenities which allows for facility standards to be designed at only 60% of the national standard. This risks reinforcing a lower quality of life on this historically poorly served community and environment, down-wind of the industrial pollution and waste depots of the past. The vogue for higher density, mixed-use development featured in loft-style apartment living and café culture elsewhere in the city,[39] is translated for this multicultural east end community into a reality of overcrowding and under-provision of key amenities such as education and health, with over 600 households facing eviction to make way for the Olympic development. As Kunzman succinctly put it: 'Each story of regeneration begins with poetry and ends with real estate'.[40]

Architectural statement and form over function and the vernacular is therefore a compromise which State and cultural institutions are willing to make, despite the 'danger that the cultural status symbol can shift the emphasis onto the building and its symbolic meaning to a degree to where what is inside hardly seems to matter at all'.[41] The same could be said of the branded city as a whole, and the annual or all-year-round festival city (e.g. Montreal, New York, Barcelona, Glasgow) and international trade exhibitions which now spread out from their main fair sites (e.g. Milan, Edinburgh, Frankfurt). This risk is mitigated however not only through the sheer numbers of visitors (especially overseas tourists) and consequent attention these event-based brands generate, but by the ancillary spending through bookshops, souvenir stores, restaurants and franchises, which are not a prerequisite of exhibition entry. In fact if the visitors to these larger and themed blockbuster venues all attended the actual shows, their carrying capacity would be breached – 'passing trade' is an economic and operational necessity. Size therefore matters in cultural city economics, particularly where officially driven by cultural tourism and access objectives.

Gentrification Effects

The globalisation of populated tourist space is a phenomenon common to major cities and urban heritage sites, which Edensor has identified not only as a recent development, but one which can be understood as an expansion of inscribing power through the materialisation of bourgeois ideologies since the nineteenth century. Accordingly, places are now conceived not as nuclei of cultural belonging, foci of attachment or concern, but according to Kearns and Philo as 'bundles of social and

39   G.L. Evans, 'Mixed-Use and Urban Sustainability', *Planning in London*, No. 52 (2005b): 43–46.

40   K. Kunzman, 'Keynote Speech to Interreg III Mid-Term Conference, Lille', in *Regeneration and Renewal* (London, 19 November 2004), p. 2.

41   K. Schubert, *The Curator's Egg. The evolution of the museum concept from the French Revolution to the present day* (London: One-Off Press/Christie's, 2002), p. 98.

Figure 11.3 MACBA, Barcelona

economic opportunity competing against one another in the open market for a share of the capital investment cake'.[42] The other process which the cultural flagship brings in its wake, is therefore that of gentrification. When the Pompidou was opened, the *Les Halles* neighbourhood and its residential profile changed: 'Not so much a renewal of the old community but of displacement and wholesale gentrification'.[43] This is apparent in the areas undergoing arts and entertainment makeovers and sites of cultural flagships. At the new Tate Modern Gallery in London, receiving over 5 million visitors in its first year (but like the Pompidou Centre, many were non-paying

42   G. Kearns and C. Philo, *The City as Cultural Capital; Past and Present* (Oxford: Pergamon, 1993), p. 18.

43   Schubert, *The Curator's Egg. The evolution of the museum concept from the French Revolution to the present day*, p. 98.

visitors rather than ticket-buying gallery attenders), adjoining artists studios were lost to cultural production, as landlords and council home tenants exercising their 'right to buy' capitalised on the 'hope value' provided by this new attraction. Meanwhile up river, the original Tate Britain Gallery has seen its attendances fall dramatically, a possible zero sum game. The hinterlands behind Tate Modern meanwhile continue to contain some of the poorest areas of deprivation in London. Like the MuseumQuarter, Vienna; MACBA, Barcelona; Guggenhaim Bilbao and Pompidou, Paris, local residents 'exclude themselves' from these institutions, playing no part in their location and employment opportunities, let alone the 'culture' on offer.

The planning opportunity and branding imperative is most evident in cities undergoing radical change. In re-branded, reunified Berlin, old and new architecture is punctuated by interstitial spaces, left over from the divided city and bomb sites from the War. The Berlin Senat requires that emblems of occupation and strife are erased ('critical reconstruction'), whilst contemporary design seek to create modern façades and extensions to heritage buildings. The Reichstag now hosts Foster's new parliament building, a glass and steel structure reaching out from the roof, topped off with a designer restaurant with views over this developing city, whilst the newly opened Academy of Arts on the Pariser Platz likewise blends the modern with the historic old building which was divided by the Berlin Wall itself. The most symbolic construction however is Pottsdamer Platz, masterplanned by Renzo Piano and with flagship buildings by Richard Rogers (both designers of Pompidou, Paris). This corporate branding, emulating Times Square, New York gives prominence to global corporations MTV and Sony and otherwise ubiquitous shopping arcades. As the imported conductor of the Berlin Philharmonic Orchestra enthused: 'Berlin is 60% German, 35% New York and 5% jungle'.[44]

In a radically repositioning city such as Berlin, as with other reconstructing cities, governance and community are unstable and fluid – Berlin has lost one million inhabitants since 1990, some seeking work elsewhere (an unemployment rate of 20% – 45% among the city's Turkish community), others leaving for the suburbs as urban quality ironically deteriorates, in contrast to the flagship property developments. On the other hand a similar number (1m) of new migrants from central and eastern Europe, and international entrepreneurs, students and artists are attracted by cheap property rents, free higher education and low property taxes. A similar number are aged 65 or over, placing strains on an already crippled public purse. Berlin's state incentive-driven property boom, drawing the city into fiscal crisis, now has several thousand square meters of unoccupied apartments and oversupply of office space (a ratio of 4 to 1 supply-demand). Berlin has no comprehensive city plan, but brands itself variously as *Life Science Capital, Music Capital, Capital of Talent* and 'City of Promise and New Beginning'. However from one Berliner's viewpoint, it is a 'City of Missed Opportunities' typified by a decline in social behaviour and neglect.[45]

---

44 BBDC, *Music in Berlin*, (Berlin: Berlin Business Development Corporation, 2005).

45 W. Nowak, *The Urban Age project.* Presentation at the Canadian University Centre, Berlin, 24 June 2005.

Figure 11.4   Reichstag and Pottsdamer Platz, Berlin

Conclusion

The hard branding of the culture city therefore looks to the power and practices of commercial branding and its packaged entertainment and emporiums. By so doing they benefit from the democratisation associated with leisure shopping (and capture, literally, these same shoppers) and the urban consumption experience. Jackson in particular suggests that the perjorative public-private dialectic may be unhelpful, and that: 'notions of consumer citizenship need to be carefully situated in and socially differentiated ... Rather than assuming that commodification and privatisation are inherently undemocratic and reactionary social processes'.[46] Balibrea on the other hand, writing from the exemplar culture capital Barcelona,[47] still sees in this situation, 'the erosion of the 'public' where redefined space ... is more and more radically commodified and dependent on private producers deciding who has access to their – also private – spaces of consumption'.[48] Barcelona's urban growth challenge, with further expansion from the old and original new city, may also signal the end to its heyday as existing residents look to the suburbs to escape from the intense inner urban pressures and onward march of gentrification, and as poorer migrants replace them – immigration has increased from 2% to 14% with 30% of the old city occupied by new migrants from Latin America, North Africa and Pakistan, whilst the city had previously attracted domestic economic migrants from poorer areas of Spain, such as Galicia and Andalusia. The city's population last peaked in 1979, and the suburban drift continues, taking decanted social housing and middle class residents alike. Like Berlin this changing demographic profile challenges planning horizons and the brand identity, as the Catalonian proportion of the city population decreases. New development, post-Olympics has focused on the the the *Diagonal Mar* and Poblenou former industrial, working class district, rebranded '22@rtist'. Over 200 hectares of industrial land promises 100,000 jobs, housed in 220,000 m^2 of new facilities. The archaeology of this industrial heritage district will however be lost, whilst facilities for independent artists in this digital media city are surprisingly limited, with its prime stakeholders, a private sector company MediaPro, a relocating University of the Arts (Pompeu Fabra) and the City.[49] Another problem with this digital media district phenomenon, is the *fact* (if not *awareness* by their proponents) that these are fast multiplying worldwide – from Malaysia (Digital Media corridor), Singapore (Media City) and Seoul (Digital City), to Berlin (Silicon Allee), Dublin

---

46   P. Jackson, 'Domesticating the street', in N.R. Fyfe (ed.), *Images of the Street: Planning, Identity and Control in Public Space* (London: Routledge, 1998), p. 188.

47   Monclus, F.-J. 'The Barcelona model: an original formula? From 'reconstruction' to strategic urban projects', *Planning Perspectives*, 18 (2003): pp. 399–421.

48   M.P. Balibrea, 'Urbanism, culture and the post-industrial city: challenging the "Barcelona model"', *Journal of Spanish Cultural Studies*, 2/2 (2001): pp. 187–210, p.195.

49   C. Gdaniec, 'Cultural Industries, information technology and the regeneration of post-industrial urban landscapes. Poblenou in Barcelona – a virtual city?', *GeoJournal*, 50 (2000): pp. 379–387; V. Lorenz, *Presentation: Centre d'Inovacio Barcelona Media*, Poblenou, Barcelona, 22 June 2005.

Figure 11.5   Hangar and Poblenou, Barcelona

('Hub'), New York (Silicon Alley) and Dundee – to name only a few. As Scott already pointed out: 'the experience of many local economic development efforts over the 1980s demonstrates, it is in general not advisable to attempt to become a Silicon Valley when Silicon Valley exists elsewhere'.[50]

New cultural facilities also suffer the fate of brand decay and the high maintenance (i.e. reinvestment) required to retain market share. Cities such as Bilbao will shortly be facing this reality, whilst the culture-led regeneration of cities such as Barcelona[51] and Glasgow[52] continue to pursue growth and investment in their cultural brands and urban infrastructures, as central to their economic development strategies. Success in both cases is measured by inward investment and tourism, rather than endogenous growth or sustainable development goals. Since substantial public resources are required, sums and levels of risk which few private investors would countenance, their opportunity cost is also high. High because of the scale and scope which political and cultural proponents require in order to assuage resistance to such mega-projects. In previous eras such resistance was muted or minimal (e.g. World Fairs, public museums).

The cost which is becoming increasingly evident and common, irrespective of their location or culture, is borne in terms of cultural diversity and production (versus consumption and mediation); in community cultural activity and amenity; and by those who do not have a stake in the gentrification processes which attach to these emerging cultural quarters: 'what is being branded in these cities is not just the immediate institution, or anything so arcane as a collection, but the city itself'.[53] However, cities have enduring cultural orientations which exist and function relatively independent of their populations at any particular time, as Lee suggests:

> in this sense we can describe a city as having a certain cultural character ... which clearly transcends the popular representations of the populations of certain cities, or that manifestly expressed by a city's public and private institutions.[54]

This is a natural tension, between tradition and heritage, continuity and change, and which city planning seeks to mediate and at its best inspire. If there is something symbolic, even metaphysical at play, commercial and competitive city cultural branding may produce uneven, unpredictable results.

---

50 A. Scott, *The Cultural Economy of Cities* (New York: Sage, 2000), p. 27.

51 O. Ballaguer, *Barcelona, Culture on the Move* (Barcelona: Institute of Culture of Barcelona, Ajuntament de Barcelona, June 2005).

52 GCC, *Arts and Creative Industries Strategy for Glasgow* (Glasgow: Glasgow City Council, 2005) (unpublished).

53 R. Ryan, 'New frontiers', in *Tate Modern special issue*, 21, London (2000): 90–96, p.91.

54 M. Lee, 'Relocating location: cultural geography, the specificity of place and the City of Habitus', in J. McGuigan (ed.), *Cultural Methodologies* (London: Sage, 1997), p.132.

Chapter 12

International Exhibitions and Planning. Hosting Large-scale Events as Place Promotion and as Catalysts of Urban Regeneration

F. Javier Monclús

F. Javier Monclús Professor of Urbanism at the Polytechnical University of Catalonia, Barcelona (Spain), he was the Convenor of the 11th IPHS Conference (2004) and is now working as a Planner for the *Consortium Zaragoza Expo 2008*. He is the co-editor of the journal *Perspectivas Urbanas /Urban Perspectives*. Among his books are the *Atlas histórico de ciudades europeas* (co-ed. 1994-96) and *La ciudad dispersa* (1998). His main interests are concerned with expos, urban strategic projects and planning models, and he has numerous publications in the area of urbanism and planning history.

Introduction

Large-scale international events and their associated urban strategies have become influential elements in the reconversion of planning over the last years. Spectacular, emblematic or 'flagship projects' linked to those events can be seen not only as part of place promotion and city marketing strategies, but also as catalysts of urban regeneration. Explicit branding policies related to these 'event projects' are not so different to those which are applied in other cases in which there is not a direct link to special events. The analysis of the planning experiences associated to the large events constitutes an interesting area of research given that these can be seen not only as temporary spatial 'hubs' or 'socio-temporal nodes',[1] but also as episodes in which the architectural and planning culture of each moment is expressed, in the specific conditions of the host cities.

---

1   M. Roche, *Mega-Events and Modernity: Olympics and expos in the growth of global culture* (London: Routledge, 2000), pp. 233–234.

Hosting large-scale events such as International Exhibitions, World's Fairs or *expos*[2] has been seen as an expression of the power of nations, celebrations of progress and the advances of industry and technology. For many scholars researching the expos phenomenon, these events can be understood as episodes belonging to the 'city of the industrial era', suggesting a supposed anachronism of their persistence in a world dominated by instantaneous media. In Walter Benjamin's vision, for example, 'World Exhibitions were sites of pilgrimages to the fetish Commodity' emphasising their ephemeral character.[3] For many authors the world of the *expos* is the paradigm of the 'ephemeral', of the temporary spectacle, of the festivals and temporary fairs.[4] In the wide literature on the theme, the *expos* have tended to be seen as an 'expression' or 'reflection' of the different political, economic and cultural contexts, much more than as agents of change or as conscious strategies aimed at transforming the host cities.[5]

Contrasting with this emphasis on the ephemeral nature of the *expos* and the lack of historic analysis of their role in urban planning and urban development, 'Cultural Planning', 'City Branding' and the strategic projects associated to the *expos* are understood by other authors as a response to the new neo-liberal context and the competitiveness between cities, characteristic of an increasingly globalised economy.[6] At the same time, from the point of view of the sectors implicated more or less directly in the *expos*, they can be seen as important instruments of renewal and as 'catalysts' of urban regeneration with substantial, increasing and lasting physical impacts.[7]

The architecture and, in part, the urban layouts of the *expos* have been the focus of a considerable number of studies.[8] Mike Sorkin points to the eventual innovative

---

2   Following M. Roche criteria, the term *expo* will be used to refer to International Exhibitions (U.K.), World's Fairs (USA), expositions, etc. M. Roche, *op. cit.*, p.1.

3   W. Benjamin, 'Paris – the capital of the nineteenth century, III. Grandville or the World Exhibitions', in *Charles Baudelaire: a Lyric Poet in the Era of High Capitalism* (1935) (London: New Left Books, 1973), pp. 164–6.

4   P. Greenhalgh, *Ephemeral vistas : the expositions universelles, great exhibitions and world's fairs,1851–1939* (Manchester: Manchester University Press, 1988); and D. Canogar, *Ciudades efímeras. exposiciones Universales: Espectáculo y Tecnología* (Madrid: Julio Ollero editor, 1992).

5   An excellent resource/reference for most fairs up until 1988: J.E. Findling, K. Pelle (eds), *Historical Dictionary of World's Fairs and Expositions, 1851–1988* (Westport, Connecticut: Greenwood Press, 1990). Also: B. Schroeder, A. Rasmussen, *Les fastes du progès. Le guide des Expositions universelles 1851–1992* (París: Flammarion, 1992). A recent global review of Olympic Games and Expos: M. Meyer-Kunzel, *Der planbare Nutzen. Stadtentwicklung durch Weltausstellungen und Olympische Spiele* (Hamburg: Dölling und Galitz, 2000).

6   G. Evans, *Cultural Planning. An urban renaissance?* (London: Routledge, 2001).

7   Bureau International des Expositions, *Les effects durables de l'éphémère. Le rôle des expositions internationales dans la transformation de la ville*, (París: B.I.E., 2001).

8   J. Allwood, *The Great Exhibitions* (London: Studio Vista, 1977); and E. Matie, *World's Fairs* (New York: Princeton Architectural Press, 1998).

role of the *expos* in the planning and architecture of the enclosures, proposing them as formal models of modern cities.[9] However, the analysis of the *expos* as place promotion strategies and as catalysts of urban regeneration has not been the subject of in-depth specific analysis, other than monographs which have examined isolated case studies.[10] Only some scholars have been interested in the socioeconomic impact of the *expos* and almost none in the aspects concerning urban development and planning.[11] In general, local interpretations predominate: one would be able to think about each Expo as a unique episode that could only be understood as part of the each city's urban history or 'micro-history'. Nevertheless, as Marcel Galopin indicates, 'it is evident that they have had, in space and the time, interaction and mutual enrichment, although it is true that each Expo carries the "mark" of the host country. It is through the reinterpretation and adaptation of previous 'models' that the *expos* become places and occasions of innovation.'[12]

In order to understand the role of *expos* with a planning perspective we should consider them as a genre and a phenomenon which has a relative historic coherence, in spite of the evident cultural and geographical diversity of the host cities. The approach taken in this study is to examine *expos* with a long term view, with a timing that seems adjusted to the successive conceptions and dominant processes in three large historic periods: the 'historical' *expos*, the *expos* of the 'era of modernity' and the recent *expos* corresponding to the period which we refer as the 'era of globalisation'. As opposed to the generalised attention to the first period, our focus here will be in the *expos* of the two later periods. Of the approximately 50 events considered, more than half (30) correspond to the ones that were celebrated from the 1930s to the present time and have been directly analysed. It allows us to observe innovations and the continuity of *expos* and their associated urban projects in the different places and historic moments considered.[13]

---

9   '... the fairs also became models, adopted visionary urbanism as an aspect of their agendas, both offering themselves as models of urban organisation and providing within their pavilions, panoramic visions of even more advanced cities to come': M. Sorkin, 'See You in Disneyland', in M. Sorkin (ed.), *Variations on a theme park. The new American city and the end of public space* (New York: Hill and Wang, 1992), p. 210.

10   To date, no study concerning the impact of the expos exists which is comparable to that by Chalkley and Essex: on the Olympic Games: B. Chalkley and S. Essex, 'Urban development through hosting international events: a history of the Olympic Games', *Planning Perspectives*, 14/4 (1999): 369–394. The research referred to in this chapter forms part of a more extensive examination of planning and international expositions entitled 'Urbanismo de las Exposiciones Internacionales' (forthcoming, in Spanish).

11   C. Servant and I. Takeda, *Study on the impact of International expositions* (París: B.I.E., 1996); and J.R. Mullin, *World's Fairs and Their Impact Upon Urban Planning* (Monticello, Ill.: Council of Planning Librarians, Exchange Bibliography, 303, 1972).

12   M. Galopin, *Les expositions internationales au XXe siècle et le bureau international des expositions* (París: Harmattan, 1997), pp. 9–10.

13   There is no agreement on the number of expositions considered internationally. Allwood identified 17 'major events': J. Allwood, *The Great Exhibitions.* J. Gold and M.

The principal objective of the research is to analyse the genealogy and the characteristics of the current 'policies of image' and urban regeneration that began in the nineteenth century, but which have been brought up to date and progressively reconverted over the last years. From this analysis, a number of interesting lessons can be learned concerning the changing nature, the limits and possible opportunities of such events, considered as forms of planning intervention, producing differing physical impacts in the host cities.

Ephemeral Events with Unplanned Physical Impacts. The 'Historical' *Expos* (1851–1929)

The vision of the world of *expos* as sorts of basically ephemeral events – reflecting the development of capitalism, nationalism and imperialism – has been widely referred to, especially regarding the so-called 'historical' ones, that is those of the second half of the nineteenth century and the first third of the twentieth century.[14] Nevertheless, this idea seems too generic from our perspective, as it is evident that the *expos* have not had the same impact on the different host cities. First of all, we need to consider the different importance (in visits, but above all in site coverage of each one of the *expos*) (see Table 12.1). Although it is not the only indicator – neither is that of the usual balance sheet – the number of counted 'visits' seems the safest one. The site areas involved are very different: from the 10–15 ha of the first *expos* of London and Paris to more than 500 ha in St. Louis or San Francisco. If one keeps in mind these two variables, the individual circumstances of each city, and their respective site coverage, we can understand that we are analysing events which are a lot more diverse than one would at first imagine.

Still more important is the fact that two *expos* of a similar size can have very different inputs and relevance. This situation occurred with the case of the first two *expos* in London during the nineteenth century. The historiography of *expos* has dedicated so much attention to the Great Exhibition of London of 1851 that it has generally neglected that of 1862, one of those that had a greater impact if we take

---

Gold affirm that up until the Second World War just 35 events merited the description as 'international exposition' (J. Gold and M. Gold, *Cities of Culture. Staging International Festivals and the Urban Agenda* (Aldershot: Ashgate, 2005), p.77). By contrast, according to Rydell-Gwinn, 'By the First World War nearly one hundred international exhibitions had been held around the world': R.W. Rydell and N. Gwinn (eds), *Fair Representations: World's Fairs and the Modern World* (Amsterdam: VU University Press, 1994), p. 3. John Findling and Kimberly Pelle find 96 fairs worthy of treatment in the text and identify 211 also-rans: J. Findling, K. Pelle, *Historical Dictionary of World's Fairs and expositions, 1851–1988* (Westport, Connecticut: Greenwood Press, 1990).

14   The term 'historical' is used by the B.I.E. referring the expos before its creation in 1928. This criterion will be adopted to clearly distinguish the expos of the 'first modernity', from the 'modern' and 'postmodern' ones.

a long-term view.[15] Certainly, the Great Exhibition hardly left any physical trace: the Crystal Palace that was rebuilt in another place would have been the legacy of that kind of 'ephemeral installation' in the landscape of Hyde Park, disappeared in a fire in 1930. On the other hand, the exposition of 1862 led to the creation of a very important 'cultural pole': the museum area of Kensington (Victoria and Albert Museum, Natural History Museum, Science Museum, etc.). The purchase of land with the benefits obtained from the Great Exhibition of 1851 allowed for its subsequent management according to the organisers' intentions to create a nucleus in this area, dedicated to the 'teaching and to the diffusion of science'. It is important to emphasise that all this occurred in the years following the Expo (in 1857 Marlborough House was relocated but became the Victoria and Albert Museum in 1909, the Royal Albert Hall was inaugurated in 1871, and the Imperial Institute was founded in 1887, etc).[16] If evaluated with the traditional criteria, the Expo of 1862 would not have been that different from its predecessor (with a somewhat lower number of visits and marginally larger site coverage). For that reason the urban impact of this Expo, as of others that will now be referred to, can be understood in different ways.

Concerning the Parisian *expos*, D. Harvey has underlined their role as enormous celebrations of modern technologies and offering a considerable stimulus to the economy.[17] Maurice Garden indicates that the *expos* became 'gigantic urban operations that for a great deal of the time shaped the west of Paris, Chaillot and the Trocadero on the right bank of the Seine to the Champ de Mars on the left bank'. As in the case of London, it is a matter of this consolidated western area being the representative and most qualified sector of the city: the 'Beaux quartiers' in which the residence of the urban elites was built.[18] It is important to comment upon the continuity in the sites chosen during all this period. As opposed to those of London, the *expos* staged in Paris (1855, 1867, 1878, 1889, 1900, 1925 and 1937) were not located in an existing park but in the western central sector of the city, a relatively consolidated urban area, between Champ de Mars-Invalides and Chaillot (the exception was the Colonial exposition in 1931 which was located in Vincennes). With this situation of relative urban centrality, it is not surprising to find a direct link between the organisation of the *expos* and the growth and shape of those new sectors of the city.

---

15   London exhibitions have been widely analysed. Besides the works by Allwood, Rydell and others referred to previously in endnote 10, some recent work which analyses the 1851 Exhibitions could be seen: P. Van Wesemael, *Architecture of Instruction and Delight. A socio-historical analysis of World Exhibitions as a didactic phenomenon (1798–1851–1970)* (Rotterdam: 010 Publishers, 2001); and J. Gold and M. Gold, *Cities of Culture*, pp. 49–76

16   J. Davis, 'The Great Exhibition of 1851. The urban dimension', in Bureau International des Expositions, *Les effects durables de l'éphémère. Le rôle des expositions internationales dans la transformation de la ville* (París: B.I.E., 2001). See also J. Gold and M. Gold, *Cities of Culture*, pp. 73–73.

17   D. Harvey, *Paris, Capital of Modernity* (New York and London: Routledge, 2003).

18   M. Garden, 'París', in *Atlas histórico de ciudades europeas. Francia* (Barcelona: Salvat-Hachette, 1996), pp. 46–47.

Table 12.1 'Historical' expos and their impacts

	Visits (mil.)	Size (ha)	Site and previous uses	Post Expo uses Legacy	Impact on urban structure and 'catalyst' effects
London 1851	6.0	10.4	Hyde Park (Central W)	Crystal Palace	marginal
Paris 1855	5.16	15.2	Cours la Reine (Central W)		public spaces improvement
London 1862	6.1	12.15	Kensington Gard. (Central W)	Museums A. and Victoria NHM, SM	Kensington Museums
Paris 1867	15	68.7	Champ de Mars (Central W)		public spaces improvement
Vienna 1873	7.2	233	Prater		Park and Waterfront i.
Philadelphia 1876	10	115	Fairmount Park (NW Periphery)		Park and Waterfront i.
Paris 1878–79	16.1	75	Champ de Mars/Chaillot/Invalids	Palais Trocadero	public spaces improvement
Glasgow 1888–89	5.7	28	Kelvingrove Park	Museums	Park
Barcelona 1888	1.2	46.5	Ciudadela Park (NE Central)	Museums	Park
Paris 1889	32.2	96	Champ de Mars/Chaillot	Eiffel Tower	Infrastructures (underground)
Chicago 1893	27.5	290	Jackson Park (10 km to CBD)	Museum of Science and Ind.	Park

Paris 1900	50.8	50,8	Champ de Mars/Chaillot	Grand et Petit Palais/Pont Alex.III	Museums area (underground)
St. Louis 1904	19.6	500	W periphery (Forest Park)		Park
Liege 1905	7.0	70			
Zaragoza 1908	0.5	10	Huerta de Sta. Engracia	Public buildings	Central extension
Gand 1913	9.5	130			
San Francisco 1915	19	635	Harbor View 4 km waterfront	Palace of Fine Arts	Waterfront regeneration
London 1924–25	27.1	86.4	Wembley	W. Stadium	Park + south extension
Seville 1929	0.8	134	Central south	Plaza de España	
Barcelona 1929–30	1.73/ 2.5	118	Montjuïc Hill (2 km W CBD)	Museums	Park + infrastructures (underground)

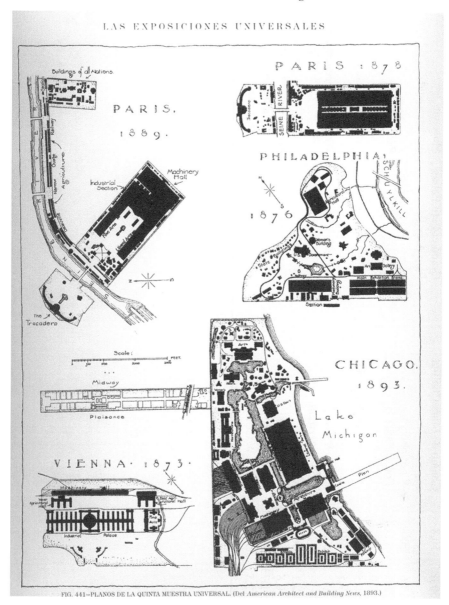

LAS EXPOSICIONES UNIVERSALES

FIG. 441–PLANOS DE LA QUINTA MUESTRA UNIVERSAL. (Del *American Architect and Building News*, 1893.)

Figure 12.1   19th Century Universal Expos: París 1889, Vienna 1873, Philadelphia 1876, Paris 1878 and Chicago 1893.

Sources: Werner Hegemann and Elbert Peets, *The American Vitruvius*, 1922 W. Hegemann, *The American Vitruvius: an Architects' Handbook of Civic Art* (1922). (Spanish version: *Arte Civil*, *Caja de Arquitectos*, Madrid, 1993).

According to that early urban implication, and in spite of the character of models that the Paris *expos* had on other cities, one must emphasise the exceptionality of the case of Paris relating to other host cities which, generally, chose sites in existing parks (or in parks in the process of being created), with the vague will to promote their consolidation. For instance, in the case of Vienna, the 1873 Expo was not included in the large construction project of the Ring, although its preparation coincided with that great urban operation. The embellishment of Vienna did not realise the capacity of the Expo, and its location in a park beside the Danube (Prater) better reveals its conception as part of the city's park system. In the North American *expos*, the preference for peripheral locations seems clear. This was the case of Philadelphia in 1876, when a park situated a certain distance from the centre of the city was chosen. The city's elites managed to obtain the right to use the Fairmount Park, the major 'urban park' of the country. On the other hand, the idea of setting six large buildings, instead of the unitary Building (as a main Pavilion) of the former *expos*, favoured the decision to give a permanent character to one of them, the Memorial Hall, dedicated to a great exposition of art. Anyway, when the men of business began to think about the organisation of the event, they were more concerned in showing the progress reached by the United States after the Civil War[19] than in urban development issues. The case of the famous 1893 World's Columbian Exposition has been extensively analysed by architectural and planning historiography, given the influence that the construction of the so-called 'White City' had in architecture and, above all, in the 'City Beautiful' movement in the United States.[20] The decision to choose a park, then in formation, at a considerable distance from the urban centre, was the object of much discussion, considering the advantages and disadvantages of each one of the possible locations. However, the physical impact of the Fair was not extraordinary, regarding the size and dynamism of the metropolis at the end of the nineteenth century.[21]

---

19   R.W. Rydell, *All the world's a fair, visions of empire at the American international expositions, 1876–1916* (Chicago-Londres: University of Chicago Press, 1984); and R.W. Rydell, J.E. Findling, K. Pelle, *Fair America. World's Fairs in the United States* (Washington and London: Smithsonian Institution Press, 2000).

20   H.H. Bancroft, *The book of the Fair* (Chicago, San Francisco: The Bancroft Company, 1893); R.W. Rydell, *The Books of the Fairs: A Guide to World's Fair Historiography Chicago* (Chicago: American Library Association, 1992); J.E. Findling, *Chicago's great world's fairs* (Manchester: Manchester University Press, 1994); and M. Manieri Elia, 'Por una ciudad imperial', in G. Ciucci, F.Dal Co, M. Manieri-Elia, M. Tafuri, *La ciudad americana* (Barcelona: Gustavo Gili, 1975). See also a comparative approach in W.Hegemann, *The American Vitruvius: an Architects' Handbook of Civic Art* (1922) (New York: Princeton A. P., 1988).

21   Even though some authors such as R.W. Rydell highlight its importance in the transformation of urban space: 'a massive reclamation effort and a spectacular urban park for the city of Chicago'. R.W. Rydell, 'Life buoy or concrete jacket? The impact of world's fairs on Chicago', in Bureau International des Expositions, *Les effects durables de l'éphémère. Le rôle des expositions internationales dans la transformation de la ville* (París: B.I.E., 2001).

The influence of the Chicago Fair 'model' in the subsequent organisation of events was important, above all because of its economic and popular success. Both the exposition of St. Louis in 1901, as well as that of San Francisco in 1915 followed the general guidelines of Chicago. In St. Louis a peripheral site by the riverside was chosen, the Forest Park. As in Chicago, the organisers did not seem to have any view of the role that the Fair could have played in urban development. By contrast, in San Francisco an ambitious urban proposal of reform can be found after the destruction of the city caused by the fire of 1906. As opposed to Chicago or St. Louis, San Francisco's strategy was based on the idea that the utilisation of the waterfront along the harbour would involve 'permanent improvements after the exposition was over', and each of the three proposals for the site selected by the exposition Board included the 'principle of permanent improvements'. Even so, that 'city of palaces' disappeared after some months 'to make way for the expansion of the city of San Francisco'.[22]

As occurred in other non-capital cities, this 'principle of permanent improvements' was present in the two *expos* organised in Barcelona in 1888 and 1929. It is important also to bear in mind that the size and importance of both events was quite modest with regard to others developed in that period (see Table 12.1). Analysing that of 1888, diverse authors coincide in indicating that the affirmation of the expectations of the Catalonian bourgeoisie at the time was more important than the legacy and the impact produced by the exposition on the urban structure. Even though the 1888 Exhibition was a successful one, it was not responsible for the important transformations which modernised Barcelona at the end of the nineteenth century, as some recent analysts suggest. It is true that the Exhibition had a positive effect on the construction of the Ciudadela Park, which had been progressing slowly since 1871. However, the objectives associated with a propagandistic desire to promote and renew the city's image prevailed over other structural issues.[23]

As for that of 1929, a certain consensus also exists concerning its evaluation as an event which worked as a tool and catalyst for the development of an extensive section of the Montjuïc mountain, which until then had been used for military purposes: its direct impact was on the consolidation of a new park and the building of the monumental Plaza de España. Besides that, the entire city benefited from a system of avenues, parks, buildings, and plazas built at this time, as well as from the execution of certain works of infrastructures (railway station, underground railway, etc.).[24] Nevertheless, we can still say that the planning strategy of the Expo of 1929 continued in the tradition to make use of the parks and existing open

22   B. Benedict et al., *The anthropology of World's Fairs. San Francisco's Panama International exposition of 1915* (London-Berkeley: Scolar Press, 1983), pp. 86–87.

23   A. García-Espuche, M.Guardia, F.J.Monclús, J.L.Oyón, 'Modernization and urban beautification: the 1888 Barcelona World's Fair', *Planning Perspectives*, 6/2 (1991): pp. 139–159.

24   C. Grandas, *L'Exposició internacional de Barcelona de 1929* (Barcelona: Els llibres de la frontera, 1988).

spaces, catalysing some improvements on them, without a clear strategy directed towards securing an important impact on the urban structure of the city. In any case, conceptions were slowly changing as an important Catalonian politician (F. Cambó, 1914) showed when speaking on the urban role of exhibitions: 'they finally become the pretext, the opportunity to sort out the largest number of urban problems, reforms and embellishments needed'.[25] In any case, these two Barcelona events – as with other historical *expos* – were still ephemeral *expos* which sometimes had important but unplanned physical impacts.

Urban Transformation as Prime Motivation: The *Expos* of the 'Era of High Modernity' and Years of Urban Growth

The expositions celebrated from the 1930s, in a context of a generalised profound economic depression, showed a change towards other more utilitarian visions in which the urban transformations were converted into the first objective to justify the celebration of the *expos*. In the same way as the 'historical *expos*', based on work and heavy industry became more and more obsolete – as concepts of *expos* were gradually assuming the new coordinates of the world of television and of the 'era of high modernity' – their associated planning strategies were also reformulated. It seems convincing that 'historical' *expos* constituted an end in themselves, even if they could have important urban impacts. In reality, regarding legacy, the vision was that finishing the event, the majority of the constructions should disappear, the place should be converted back to open space once more and be prepared for other uses.[26] Some occasional agreements permitted the conversion of certain buildings (palaces or symbolic monuments) and their adaptation for new uses. However the organisers were not, or at least were barely, worried about the future destiny of the sites. After the 1930s, this view was modified, especially after the Second World War, to the extent that the very legitimacy of holding *expos* would increasingly become determined by the benefits of investments. It is obvious that this did not occur overnight. In the *expos* of Chicago 1933[27] or Paris 1937,[28] that concern still occupied a relatively secondary place, and changes can be appreciated in the conceptions and in the strategies rather than in the legacies and final transformation of urban areas. The main legacies and impacts of this cycle of *expos* are indicated in Table 12.2.

---

25   Quoted by I. Solà-Morales, *L'Exposició Internacional de Barcelona 1914–1929: Arquitectura i ciutat*, (Barcelona: Fira de Barcelona, 1985).

26   Galopin, p. 123.

27   R.W. Rydell, *World of Fairs: The Century-of-Progress Expositions* (Chicago: University of Chicago Press, 1993).

28   In the official documents of the 1937 Paris Expo there is still a reference to the construction of a 'ville éphémère' ('ephemeral city'), despite the conceptual and organisational renovation manifested in the 'possibilité d'apporter des améliorations qui pourraient subsister l'Exposition disparue': Bureau International des Expositions (B.I.E.), *Exposition International des Arts et Techniques. Paris 1937. Rapport Gènéral*, pp. 13–230 (B9 A1 E3 1468).

Table 12.2 Modern *expos*

	Visits (mil.)	Size (ha)	Site and previous uses	Post Expo uses Legacy	Impact on urban structure and 'catalyst' effects
Paris 1931	32	55.2	Vincennes	Museum	marginal
Chicago 1933–34	38.8	170	Lake Michigan shores		Park Harbour
Brussels 1935			Heysel Plateau Park		
Paris 1937	31.4	105	Champ de Mars/ Chaillot	Palais de Chaillot	Museums area
New York 1939–40	44.9	500	Flushing Meadows (Queens)		Park Site clearing
New York 1964	51.6	500	Flushing Meadows (Queens)	Stadium, State Pavilion Hall of Science	Park regeneration
London 1951	8.5	11	South Bank (derelict a.)	Royal Festival Hall	South Bank regeneration
Brussels 1958	41.5	200	Heysel Plateau Park	Atomium	Heysel Exhibition Centre
Seattle 1962	5.6	30	North downtown (1 m)	Convention Center	Seattle Center
Montreal 1967	50.3	400	Riverside Islands	Biosphere, Art gallery	Infrastructures
San Antonio 1968	6.4	36.8	Riverside	Paseo del Rio	
Osaka 1970	64.2	351	Suita new town (10 km CBD)		Riverside regeneration Cultural Park
Spokane 1974	5.6	40	Spokane riverside	Convention Center	Riverside regeneration Infrastructures
Knoxville 1982	11	28.8		Convention Center	Riverside regeneration Infrastructures
Lousiana/N. Orleans 1984	7.3	32	Mississippi riverside		Riverside regeneration Science City

Tsukuba 1985	20.3	102	New Town (50 k NTokyo)	Convention Center	Waterfront regeneration
Vancouver 1986	22.1	70	South downtown South Bank	Convention Center	Theme Park
Brisbane 1988	18.5	40	Brisbane river		

**Figure 12.2   New York World's Fair of 1939. Planning the post-fair as a
         metropolitan park.**

Source: M. Meyer-Künzel, *Der planbare Nutzen. Stadtentwicklung durch Weltausstellungen
und Olympische Spiele.* (Hamburg 2000: Dölling und Galitz).

The changes in the urban development conceptions were significant in the famous
New York World Fair of 1939/40.[29] Not only because of the size of the site chosen
and the considerable figure of 45 million visits. Besides that, the characteristics of
the site in Flushing Meadows – an abandoned area, then being used as a landfill area
for waste disposal purposes – implied a significant change in the urban development
strategies. Again, new conceptions changed more rapidly than the effective
revaluation of the area. In this case, the main idea of Robert Moses was that the
benefits obtained could be used to convert Flushing Meadows into a new type of

---

29   M. Berman, *All that is Solid melts into Air: the experience of modernity* (London: the
Souvenir Press, 1983), p. 318; 'The New York World's Fair of 1939/40 marked the definitive
end of the nineteenth-century concept of world exhibitions': Van Vesemael, p. 445.

Figure 12.3   Montreal Expo of 1967. A new site on the Havre peninsula and the two islands on the river Saint-Laurent.

Source: D.Canogar, *Ciudades Efímeras. Exposiciones Universales: Espectáculo y Tecnología* (Madrid: Julio Ollero editor, 1992).

park. As this desire was not able to be realised then, Moses sought to achieve it again in 1964, organising a new exposition in the same place. Although the final balance of the transformation of the site is still the subject of debate, it seems clear that the planning strategy played a leading role in relation to the other objectives of an economic, political or cultural type. The slogan of the Expo –'Building the World of Tomorrow'– sums up well the visions that then prevailed in the planning culture, not so much for the planning layout – still heirs of the Beaux Arts – as for the application of transformation strategies of the contemporary metropolis, based upon the construction of complex infrastructure systems.[30]

Since the celebration of the New York World Fair of 1939, up until the end of the 1980s, only four large *expos* were held (with sites larger than 100 ha: Brussels 1958, Montreal 1967, Osaka 1970 and Tsukuba 1985, although only the first three of these were 'universal' or of the first category), besides that of New York 1964, which was held on the same site of the former 1939 Fair. In the four large *expos*

30 M. Dickstein, 'From the Thirties to the Sixties: The New York World's Fair in its own Time', in R.H. Bletter et al., *Remembering the future: the New York World's Fair from 1939 to 1964* (New York: Rizzoli, 1989).

celebrated after that of New York, up until the 1980s, different strategies of urban transformation were proposed. In the case of Brussels 1958, the chosen site was later converted into a park equipped for fairs, on the same site in which the 1935 Expo (Heysel Plateau) had been celebrated.[31] In the Montreal Expo of 1967, a singular site was chosen formed by the Havre peninsula and the two islands on the river Saint-Laurent: Saint-Hélène and Notre-Dame (the first one expanded and the second created for that occasion). It was in this Expo when a clearly utilitarian vision was imposed, in which the event served to carry out a substantial transformation of the city and of its system of collective transport.[32]

In the Expo of Osaka 1970, the option taken was for the satellite city of Suita, some 10 km north of the urban centre. This was presented as a laboratory in which the mega-structures proposed by the Japanese 'metabolist' architects could be experienced – with Kenzo Tange having responsibility for the Master Plan – as a decisive step into a new urban scale, in the line of the Montreal experience.[33] Finally, the other important Japanese exposition that of Tsukuba 1985, supposed a more radical innovation, as it was held in Tsukuba Science City, a 'new town' planned in the 1960s and located 50 km to the north of Tokyo, with an elaborate strategy of promotion of the new city as a 'world science centre'.[34] The diversity of urban contexts in those large *expos* makes it difficult to consider their respective urban 'impacts' in a generic way. For example, although the majority of the constructions of Osaka or of Tsukuba have disappeared, it does not mean that the urban impact was not prominent. The construction of infrastructures (such as the 'bullet train ' from Tokyo to Osaka) or the consolidation of a high technology research centre (as in the case of Tsukuba) can be considered as part of a process induced, at least in part, by the respective *expos*.[35]

As opposed to the large *expos*, more related to the national visions, one of the most notable features of the 'smaller' or 'specialised' ones (with a site coverage of less than 100 ha), is the greater attention given to the local interests of the host city. Instead of adopting the large *expos'* model, here specific strategies were proposed and, very often, in a more innovative way than in the large *expos* which were held in the same period. Although not an International exposition, the so-called 'Festival of Britain' held in London in 1951, was an extremely significant episode from a planning and architectural perspective. The urban operation was developed on the South Bank, a relatively central site by the River Thames, but a former marsh zone

---

31   However, the official documents do not develop the global Planning Strategies as much as the Site Planning, without disregarding the functional questions: Bureau International des Expositions (B.I.E.) (B10 A1 E2 1539).

32   Y. Jasmin, *La petite histoire d'Expo 67: L'Expo 67 comme vous ne l'avez jamais vue* (Montreal: Éditions Québec/Amerique, 1997), pp. 32–33.

33   R. Banham, *Megastructure. Urban futures of the recent past* (London: Thames and Hudson, 1976), p.103.

34   B.I.E., *Expo'75 Official Report* (B A1 E3 1791).

35   C. Servant and I. Takeda, *Study on the impact of International Expositions* (París: B.I.E., 1996).

which was derelict at the time. As happened in the 1862 Exhibition, the impact of the Festival of 1951 was marginal in the years immediately following, but needs to be seen in the long-term. Although the Royal Festival Hall was the only part that was not dismantled, today the South Bank and its extension toward the east in Bankside has became the most successful cultural pole of London, although the decisive impulse has been that of recent years. The beginning of all those operations can be situated in the unitary conception of an urban piece of not more than 11 ha that was a key for starting the regeneration of the South Bank of the River Thames.[36]

When the North American smaller *expos* of the 1960s to 1980s are analysed, it is interesting to confirm the leading role of the planning motivations, almost always directed towards the 'regeneration' of the central parts of the host cities. In fact, this was the first time that urban renewal was the prime motivation for hosting an international exhibition. That was the case of the Seattle Expo of 1962, organised by businessmen with the chief objective of stimulating growth and promoting the city's downtown area. The success of this *expo* in relation to these two objectives[37] was much higher than those of the other large expositions, such as that which was celebrated two years later in New York (and which generated significant losses). The creation of a new civic centre (Seattle Center) was to serve as an inspiration or model for other cities. In this way with the *expo* of San Antonio 1968, firstly the leaders came to an agreement on the need to link the Expo with the urban renewal of the civic centre and later the adequate site was chosen (with the argument that it could also be linked to the regeneration of the river ('Paseo del Río').[38] In that of Spokane 1974 something similar happened, perhaps still with more clarity as the Expo was focused on ecological and environmental themes, taking advantage of it for the in depth regeneration of the Spokane River.[39] In the 1980s, other 'specialised' *expos* followed similar strategies: Knoxville 1982, New Orleans 1984 (with a spectacular economic failure), Vancouver 1986 (the most successful of all) and Brisbane 1988.[40] It should be said that with these types of expos – of a 'special' or 'thematic' category

---

36   M. Banham and B. Hillier (eds), *A Tonic to the Nation: The Festival of Britain 1951* (London: Thames and Hudson, 1976); B.E. Conekin, *The Autobiography of a nation: the 1951 Festival of Britain* (Manchester New York, N.Y.: Manchester University Presscop., 2003); and Allwood, *The Great Exhibitions*, pp. 151–153.

37   J.M. Findlay, *Magic Lands. Western Cityscapes and American Culture After 1940* (Berkeley and Los Angeles: California University Press, 1992).

38   The Post-Fair use of the ground as a municipal center was clearly planned from the beginning, 'according to the City and Urban Renewal approved plan': B.I.E., Hemisfair 1968. Final Report to B.I.E. (B11 A1 E2 1689).

39   J. Allwood, *The Great Exhibitions: 150 years* (London: Exhibition Consultants Ltd., 2001), pp. 147–158; R.W. Rydell, J.E. Findling and K.Pelle, *Fair America. World's Fairs in the United States,* pp. 100–130.

40   With some of them not so 'small', such as the one in Vancouver, with 70 ha and 22 millions visits. Even with a deficit of $ 300 million, this was regarded as 'a price worth paying for the various economic benefits the expo succeeded in delivering to the city and the province': Roche, *Mega-Events and Modernity*, p. 133.

– it was not the participating countries but the organisers who built the different buildings. This introduced a very significant element, as the capacity to control the architectural and urban layout enlarged considerably. Instead of being seen as a 'reduced' version of the large *expos*, it is important to think that in these smaller *expos* urban operations have been easier to integrate in the urban structure of the city, with greater possibilities of planning their subsequent reconversion, keeping in mind that their more limited surfaces can be assigned to specific uses, beyond the tradition of leaving some isolated buildings as the 'museums in the park' formula.

City Marketing and Place Promotion as a Main Goal. The Postmodern *Expos* of the 'Era of Globalisation'

If one places them in the previous historical sequence, the 'postmodern' *expos* (those celebrated from the 1990s) respond more to the strategies based upon urban promotion and marketing, than the previous events, based on the celebration of industrial and technological progress. In this sense, the modernisation experienced is notable by a genre initiated in the nineteenth century, up until the twentieth century, even though it had been renovated substantially following the Second World Ear. What seems beyond doubt is that, in spite of the affirmations of some in the sense of their possible disappearance in a world increasingly dominated by the new media of communication, not only that fact has not been seen but we can verify how cities of different sizes and geographical areas continue competing in order to organise *expos* or events of similar characteristics. It seems clear, as indicated by Maurice Roche that 'the 1990s have seen a renewed interest in these mega-events as a cultural genre and a general reactivation of the expo movement'.[41] In analysing the proposals advanced and implemented in the last 15 years we can include a series of events, six of which would correspond to the *expos* recognised by the B.I.E.: that of Paris 1989 approved but finally not carried out,[42] Seville 1992, Taejon 1993, Lisbon 1998, Hanover 2000, and the last one in Aichí 2005. Another three would have slightly different meanings as they do not involve the participation of nations: London Millennium 2000, Swiss Expo 2002, Forum of Cultures Barcelona 2004. Finally, we include the next events to be held in Monterrey (2007), Zaragoza (2008) and Shanghai (2012), the last two recognised by the B.I.E. The main figures, legacies and impacts foreseen in this cycle of *expos* are indicated in Table 12.3.

In the large *expos* of the 1990s – Seville 92 and Lisbon 98 – new policies of image coexist with the already experienced planning strategies of urban transformation. The

---

41   Roche, *Mega-Events and Modernity,* p. 132.

42   We include this failed project as it is relevant not only because of it belongs to the policies of image of Parisian 'Grand Projects', but also because of its pragmatic conception. The projects for the Expo of 1989 were oriented towards the recovery and regeneration of the banks of the River Seine (particularly the land occupied by old stores and railway installations in Berçy and Tolbiac). See M. Meline, P. Vercken, *Projets pour l'exposition universelle de 1989 a Paris: livre blanc* (Paris: Flammarioncop., 1985).

Table 12.3 Postmodern *expos* and large-scale events

	Visits (mil.)	Size (ha)	Site and previous uses	Post Expo uses Legacy	Impact on urban structure and 'catalyst' effects
Paris 1989	—	130	Seine Riverside		Riverside regeneration
Seville 1992	41	215	Cartuja island/ Guadalquivir	Offices Facilites	Infrastructures Technol. Park
Taejon 1993	14	50	Riverside + rural uses		Exhibition center Theme Park
Lisbon 1998	10	61/350	Tagus Riverside	Oceanarium Offices	Riverside regeneration
Hannover 2000	18.1	160	Trade Fair + rural uses	Offices Facilities	Housing, park Exhibition centre
London Millenium Exhibition 2000	6.5	53/77	Greenwich (derelict area)	Millennium Dome	Riverside regeneration New mixed development
Switzerland 2002	10	Neuchatel 17 Biel 18 Murten 39 Yverdon 17	Lakefront open spaces	Blur Building ('The Cloud')	marginal
Barcelona Forum 2004	3.3	50/320	Seafront/ Besos river (derelict area)	Convention Centre	Waterfront Regeneration Renewal and New development
Aichi 2005 (Nagoya)	15 (est)	173	Kaisho Forest Seto City, Nagoya, Aichi		
Monterrey 2007		114/150	Parque Fundidora		Eco-City
Zaragoza 2008		25/150	Ebro Riverside		Eco-Park
Shanghai 2010		240	Huangpu Riverside		

Sources: Allwood (1977, 2001), Benedict (1983), Greenhalgh (1990), Servant-Takeda (1996), Schroeder-Rassmussen (1992), Findling-Kimberley (2000), Evans (2001), Gold (2004), B.I.E.

Figure 12.4 Barcelona Forum 2004. A new Barcelona skyline and waterfront.
Source: Barcelona City Council

Seville Expo proposed – amongst other diverse objectives of a national character – a strategy aimed at changing the historic image of the city, towards a new postmodern version, that would not only attract tourism, but also investment for its economic lift-off based on advanced service activities. At the same time the Expo held in Seville showed how an event like this could be linked – in a medium sized city – to very ambitious operations with a view towards its urban development and modernisation. With a surface area of 215 ha and 41 million visits, the Expo was conceived as a tool to revitalise a city and an historically depressed region. The island of 'la Cartuja' in the Guadalquivir River was chosen as the strategic site, very close to the historic city centre. Transformations in this area have been really important and there is little doubt on the impact of the new infrastructure and facilities associated to the Expo: new bridges over the river, road infrastructure throughout the whole of the city, a high speed train connecting Seville with Madrid, etc. In spite of the recession in the years immediately after the celebration of the Expo, the truth is that the investment performed, not just in the precinct of the Expo, but also on the banks of the Guadalquivir and in the whole of the city, can be considered as an 'induced effect' of a bigger magnitude to that which other cities had experienced up until then. The strongest criticisms of the Expo 92 are centred in two different aspects: on the one hand, the excessive autonomy of the precinct in relation to the

Figure 12.5   Zaragoza Expo 2008. The next thematic International Expo. An urban project with the pretext of water

Source: Consortium Expoagua.

river and the historic centre of Seville;[43] and on the other hand, the complex process of reconversion of the facilities after the event. Although this last question has been the subject of intense debate, at times it is forgotten that as from 1988 an innovative study plan was elaborated – the Cartuja 93 project, directed by Manuel Castells and Peter Hall – for the reuse of the spaces and buildings of the Expo with a new generation technological park.[44] In any case, the impulse of the Expo 92 upon the urban structure of Seville was extraordinary.[45]

---

43   What seems remarkable is that the logic of responsibilities for Expo 92 favoured the expo Master Plan over the General Planning 'which the Council was simultaneously drawing up': V. Perez Escolano, 'Sevilla, ciudad contemporánea a impulso de exposiciones', in F.J. Monclús and M.Guardia (eds), *Planning Models and the Culture of Cities. 11th International Planning History Conference* (Barcelona: CCCB, 2004).

44   M. Castells and P. Hall, *Tecnópolis del mundo la formación de los complejos industriales del S. XXI,* (Madrid: Alianza DL, 1994).

45   V. Perez Escolano, 'A brief look back at Sevilla '92', in *Lisbon Expo 98. Projects* (Lisboa: ed. Blau, 1996).

With similar conceptions, but stressing the image of a careful environmental intervention, the Expo of Lisbon of 1998 was launched. Here, the renovation of the image of the city followed an outline not very different to that of the *waterfronts* developed in the 1980s and 1990s. With the explicit will to act directly upon an environment very much superior to that of the Expo precinct, the concept of an explicit 'intervention zone' appeared. In this way, a double objective was achieved. Firstly, that of assuring the self-financing of the Expo project; and secondly, extending the intervention in order to transform and regenerate the surrounding and deteriorated area. Also in this case, the impact of the associated projects to the Expo was strong, as all the riverfront was regenerated with dwellings and infrastructure, which led to an important alteration of the urban structure and of the Tagus façade.[46] Only two years later, at the Expo of Hanover 2000, the image of an 'ecological Expo' was of great importance. As would happen in nearly all the *expos* from then on, in the spirit of Agenda 21, the goals were to minimise land use by utilising existing fairgrounds and to ensure as far as possible the 100% post use of newly developed areas. Here there was an option for a metropolitan location, on land belonging to an existing Fair. It was a matter of the modernisation and enlargement of this Fair with the aim of turning one area into a business park after the expo (of the 160 ha, 90 ha correspond to land under the jurisdiction of the Fair, the other 70 ha being of agricultural use).[47] It does not seem coincidental that the change of millennium was seen by other cities as an opportunity to launch large projects based on amenities which were conceived as spectacular buildings, at times with low benefits and with serious problems of reutilisation. The case of the Millennium Dome in London, built as part of the Millennium Exhibition is perhaps the best-known, although not the only one. However, even in episodes such as this, it is important to avoid its interpretation as a remote and exceptional example of an isolated intervention. Rather it is verified that they form part of a general strategy of regeneration of abandoned areas – in the East of London, by the River Thames – that pass from obsolete industrial uses to harbour new service sector and residential uses. It remains to be seen if this kind of urban project can emulate the impact of the 1862 and 1951 London Exhibitions, with the shaping of new 'cultural poles'. As in other postmodern *expos* in the London Millennium projects of prestige – directed towards changing the image of the place – coexisted with other less visible and spectacular actuations, associated to transport infrastructure and urban regeneration.[48] Contrasting with these episodes, it is possible to interpret the urban and landscape strategy behind the Expo of Switzerland 2002, celebrated in four cities in a simultaneous way as exceptional. Although the 'cultural'

---

46   N. Portas, 'O pós-Expo e o resto à volta', in V. M. Ferreira and F. Indovina (org.), *A cidade da Expo '98. Una Reconversao na Frente Ribeirinha de Lisboa?* (Lisboa: Bizancio, 1999).

47   B. Breuel et al., *Expo Architektur Dokumente/Architecture Documents, Beiträge zur Weltausstellung EXPO 2000 in Hannover/Contributions for the World Exposition EXPO 2000 in Hannover* (Verlag: Hatje Cantz, 2000)

48   J. Gold and M. Gold, *Cities of Culture*, pp. 265–270.

conception of the Expo could be seen following the tradition of other Swiss *expos* (the last one was held in 1964) and could be related easily with others – conveniently brought up to date – its explicit goals were directed to produce the minimum impact in these cities. This is something which turns out to be unusual and draws to mind the more ephemeral *expos* of the 'historical' period we have seen previously.[49] Beyond promoting the image of the cities in which the Expo was held, the idea of an event that only should leave traces in the memory of the visitors – instead of on the physical structure of cities – constitutes the more original characteristic of this approach. Paradoxically, an Expo of an ephemeral character – such as those of the 'historic' period analysed – would be one of the most postmodern for its renovated forms of articulation between architecture, landscape, electronic and audio-visual communication systems.

It is interesting to place the project of the Forum of the Cultures of Barcelona 2004, in that long series of Exhibitions (universal, international or just *expos*). The official discourse in the Forum of Cultures stated that 'expos are over' and that Forum was a new and different formula. However, we should recall that the Forum sought to organise an Expo with the accreditation of the B. I. E. but this conception was 'reconverted' towards what has been defined as a 'Festival of the ideas', a Forum on the themes of peace, on cultural diversity and on sustainable development.[50] This proposal was considerably opportunist and represented the risks of events with an excessive emphasis on the policies of image.[51] In actual fact, if the global conception of the event led to a long discussion on its 'originality' regarding other *expos* celebrated to date, planning objectives and strategies can be easily related to other urban projects which were implemented with the ones of the two last periods, particularly, of the 'smaller' specialised or thematic *expos*. The idea to promote a 'pole of centrality', a business and leisure centre, with the Convention Centre and the Forum building as permanent installations is reminiscent of North American *expos*. Besides that the Forum of Cultures would also share other characteristics of the recent *expos* of medium level as far as regeneration strategies are concerned.

Viewed from this perspective, the successive Barcelona experiences that often have been interpreted with a strictly local perspective, as 'original strategies adopted by a capital city without a state that is not able to grow in another way' could be understood in a different way.[52] Without denying either the specificity or

49 Expo 02, IMAGI-NATION. Le livre officiel d'Expo 02 (Laussane: Ed. Payot, 2002).

50 Even the themes chosen were not absolutely new: 'peace' and progress were traditional ones in several *expos*, 'cultural diversity' was the theme of San Antonio Expo 1968, and 'environment' was a central theme in both Spokane 1974 and Hanover 2000.

51 For some recent critical views, see: H. Capel, *El modelo Barcelona: un examen crítico* (Barcelona: Ed. Del Serbal, 2005); and M. Delgado, *Elogi del vianant. Del 'model Barcelona' a la Barcelona real* (Barcelona: Edicions de 1984, 2005).

52 F.J. Monclús, 'Barcelona's planning strategies: from 'Paris of the South' to the 'Capital of West Mediterranean'', (The European Capital City, Amsterdam), *GeoJournal* 51/1–2 (2000): pp. 57–63; F.J. Monclús, 'The Barcelona Model: an original formula? From

the creativity developed in the city, it seems more useful and appropriate to think on the capacity of some cities – as happened in Barcelona – to work in creative versions of the models that have been experienced in other cities which have given parallel answers to parallel cultural and socio-economic dynamics. These strategies should be evaluated in the same way as in other urban projects – either in Baltimore or in Bilbao, in Barcelona 2004 or in Zaragoza 2008[53] – analysing their problems and their risks, against the advantages and benefits which can be attributed to them, whether or not they be associated to ephemeral events.

Conclusions

The principal hypothesis of this analysis of the *expos* is that they belong to an historic genre and, at the same time, that they are progressively reconverted in a type of strategic urban Project with which they share many of their objectives. Thereby, in addition to the 'official' objectives, the recent *expos* are planned more and more as strategies of place promotion and instruments to boost the cities' economies and to regenerate urban areas. If at the end of the 19th and the start of the 20th Centuries, the urban projects associated to the *expos* were inscribed in the logic of the *City Beautiful* now the new context of competitiveness between cities is the starting off point. Currently, in the analyses of 'prestige' interventions, it is usual to make analogies with an extensive type of projects of diverse dimensions and complexity, such as the 'Grands Projects' in Paris, for example.[54] Although their singularity resides in the ephemeral nature of the events to which they are associated, the urban projects which the *expos* catalyse present similar characteristics and can have planning impacts of a similar nature to other projects with non-ephemeral characteristics. The possible costs and benefits (or rather, the limits and opportunities) of the current *expos* are not that different to those of other projects not associated to ephemeral events. On the side of the costs and the risks: from the 'investment overdose', that is to say, the excessive concentration of resources in a limited space, to the physical risk of the formation of enclaves or precincts poorly integrated to the urban structure or at the danger of an excessive standardisation, themeing or banalisation of the projected spaces. On the side of the benefits and the opportunities: the *expos* can be seen as places of experimentation and authentic 'laboratories of modernisation' in which the urban spaces corresponding to each historic moment and to the conditions of each host city are foreshadowed. In addition, they can be understood as a modality of

'Reconstruction' to Strategic Urban Projects (1979–2004)', *Planning Perspectives*, 18/4 (2003): pp. 399–421.

53   F.J. Monclús, 'Projets urbains, expositions et développement durable', *Symposium International. Eau et développement durable. Un projet pour l'Expo 2008,* Bureau International des Expositions (Zaragoza: Consorcio Expo, 2004).

54   N. Portas, 'La emergencia del proyecto urbano', *Perspectivas Urbanas/Urban Perspectives*, 3 (2003): pp. 15–26.

strategic planning generating consensus.[55] A type of planning and urbanism that can contribute to the renovation of the image and identity of the city, at the same time as mobilising agents and resources, catalysing new urban projects that do not have to be limited to 'prestige projects' or to emblematic and spectacular projects. Nevertheless it should be recognised that these events fulfil a similar and pronounced role in current times as those performed by the large urban actions involving buildings and public space of the past. That is to say, they are just as susceptible to be converted into mere policies of image, as in others generating considerable economic and social benefits.

---

55 In addition to supposing 'cultural and physical bridges between elites and the people': M. Roche, *Mega-Events and Modernity*, pp. 233–234.

Chapter 13

Contemporary Urban Spectacularisation

Lilian Fessler Vaz and Paola Berenstein Jacques

Lilian Fessler Vaz, architect and urbanist, professor at the Faculty of Architecture and Urbanism of the Federal University of Rio de Janeiro (Brazil). Author of *Modernidade e moradia – habitação coletiva no Rio de Janeiro* (Rio, FAPERJ/ 7 Letras, 2002), co-author of four books of the collection História dos bairros do Rio de Janeiro (Rio, Index, 1984, 1985, 1986, 1987), and other papers.

Paola Berenstein Jacques is an architect and urbanist, professor at the Faculty of Architecture of the Federal University of Bahia (Brazil). Author of: *Les favelas de Rio* (Paris, l'Harmattan, 2001); *Estética da ginga* (Rio de Janeiro: Casa da Palavra, 2001); *Esthétique des favelas* (Paris: l'Harmattan, 2003); *Maré, vida na favela* (Rio de Janeiro: Casa da Palavra, 2002) and *Apologia da deriva* (Rio de Janeiro: Casa da Palavra, 2003).

Introduction

This chapter makes an analysis of contemporary urban interventions, highlighting those of an urban-cultural nature, and concludes with a criticism of the current spectacularisation of cities. We consider that the spectacularisation of cities, or the creation of the so-called spectacle-city, is directly related to a decrease of popular participation, which we perceive through the social and cultural gentrification of the areas where spectacular mediatic urban interventions are carried out at present.

Urban-cultural Interventions

Transformations of economic and social nature bring about countless other changes. The adaptation of structures, forms and images, either through slow processes or drastic interventions, may be observed in many cities. In the industrialisation period, the city underwent a radical transformation, adapting itself to its new condition as a material production centre. At present, in the post-industrial economy, new transformations are under way: non-material production obliges cities once again to transform themselves.

With the industrial city, the need to face the new challenges, seeking to foresee, guide and control changes, led to the emergence and development of urbanism and urban planning. With the post-industrial city, new forms of intervention are spreading, guided by strategic plans and urban projects. In the case of the industrial city, the adaptation was addressed to material production. In the case of the post-industrial city, however, the adaptation is addressed to immaterial production, that is to say, to that of common non-material goods: services, information, symbols, values and aesthetics, in addition to knowledge and technology.

This new economy which, according to Peter Hall[1], ceased to be an 'informational economy' and became a 'cultural economy', and which, according to Arantes[2], finds its 'new driving force' in culture, sets new demands for cities which are radically different from those for the production-city. Culture, now an instrument of economic development, also comes to be used as a strategy of intervention for urban revitalisation.

For the present consumer society, urban areas are considered adequate if they have advanced transport and communication means, in addition to quality in residential and environmental terms and a high level of cultural and educational offerings, bearing in mind local as well as global conditioning factors. This set of qualities stems from the rivalry between cities that seek to offer the best conditions for attracting inhabitants, capitals, investments, companies and tourists. Some areas are prioritised in these urban renewal processes, such as historical centres, degraded central areas and urban voids resulting from the deindustrialisation process. Consequently, the interventions seek to reverse the harmful effects of the post-Fordist economic changes and to adapt the built-up environment to the new economy. The renewed areas should offer conditions for the production and consumption of culture, for tourism and for the symbolic economy, in the sense postulated by Zukin,[3] of the production of cultural meanings and spaces in the city.

In recent decades there have been countless urban plans, projects and interventions in which culture[4] was underscored as a principal factor. In the field of urban planning and urbanism, new terms and expressions emerged that portrayed this importance: 'cultural territories' and 'places', 'cultural districts' and 'poles', 'cultural engineering', 'cultural planning', 'planificación cultural', 'cultural regeneration', and 'culturalisation of the city'[5], among others.

---

1   P. Hall, *Cities in Civilization* (New York: Fromm International, 2001), p. 8.

2   O. Arantes, *Urbanismo em fim de linha* (São Paulo: EDUSP, 1998), p. 152.

3   S. Zukin, *The Culture of Cities* (Oxford: Blackwell, 1995).

4   A discussion of concepts of culture does not lie within the scope of this paper. In the case at hand, a mercantilised, globalised and spectacular culture is generally involved. See L.F. Vaz and P.B. Jacques, 'Reflexões sobre o uso da cultura nos processos de revitalização urbana'. Anais do IX ENANPUR (Rio de Janeiro: 2001), pp. 664–674.

5   On 'cultural engineering', see A. Haumont (1996), 'L'équipement culturel des villes', in *Les Annales de la recherche urbaine*, 70/March (1996): pp. 148–153. On 'cultural planning', see G. Evans, *Cultural Planning, an Urban Renaissance?* (London and New York, Routledge, 2001). On 'cultural and strategic planning', see Fazer cidade: Planos, Estratégias e Desígnios, in P. Brandão and A. Remesar (coord.), *O Espaço Público e a Interdisciplinaridade* (Lisbon: Centro Português de Design), pp. 90–98. On 'cultural regeneration', see M. Wansborough and A. Mageean, 'The Role of Urban Design in Cultural Regeneration' *Journal of Urban Design*,

Although the recourse to the cultural factor in the urban project may be observed in various areas of the city, here we will lend priority to those measures addressed to the consolidated areas of cities. These measures are urban interventions, that is to say, as conceptualised by Portas:[6] '... the set of programmes and projects (...) that affect the urbanised fabrics of agglomerations, whether they are old or relatively recent, bearing in mind: their functional revitalisation or restructuration (...); their architectural rehabilitation or recovery (...), and, lastly, their social and cultural reappropriation (...)'. More specifically, these are projects for urban interventions in which strategic use is made of cultural resources. Their objective is local development and they may or may not be associated with cultural policies and plans.

The interventions that seek to readapt the existing urban fabrics to new situations, have been receiving new designations in each new context, usually with the prefix 're-': renewal, restructuration, revitalisation, rehabilitation, requalification or regeneration, among others. A quick review of the times and designations involved, allows the emergence of the cultural dimension in the framework of urban interventions to be situated in a historical perspective.[7]

From the middle of the nineteenth century, the goal was the beautification and sanitisation of the industrial city, acting on the densely occupied and hardened central areas. The modernisation processes of Paris, Barcelona and Vienna became classic forms of urban intervention and expansion. In the middle of the twentieth century, 'urban renewal' based on modern ideals of rationalism and functionalism allowed the emergence and/or development of modernised densified verticalised centres. In opposition to this type of intervention, which was often implemented after reducing the existing urban fabric to a *tabula rasa*, and with the criticisms of the destruction of the built-up heritage, the breakage of the existing social links and the always present real-estate speculation as its starting point, 'urban revitalisation' or 'rehabilitation' arose in the 1960s and 1970s. This new urban practice rejected the street as a space for circulation alone, and the monotonous homogeneous urban fabrics defined according to zonings and urban coefficients; at the same time it followed up on urban composition, recovering public spaces, the typology of buildings and the urban morphology.[8] The proposals came to be patterned by urban projects, anchored in architectural culture and valuing urban design. From the 1970s and 1980s, the emphasis on public spaces, regionalism, and the concern for the built-up heritage and history, added a cultural dimension to the urban policy. The spread of the practice of preserving and revitalising

5/2, (2000): 181–197. On 'culturisation', see M. Wansborough and A. Mageean, 'The Role of Urban Design in Cultural Regeneration', *Journal of Urban Design*, 5/2, (2000): pp. 181–197; and H. Häussermann, *Grossstadt. Soziologische Stichworte* (Opladen: Leske + Budrich, 2000), amongst others.

6   Nuno Portas, 'L'emergenza del progetto urbano', *Urbanistica*, 110/June (1998): p. 55.

7   In this respect see Vaz and Jacques, 'Reflexões sobre o uso da cultura nos processos de revitalização urbana'.

8   Françoise Choay and Pierre Merlin, *Dictionnaire d'urbanisme et de l'aménagement* (Paris: Presses Universitaires de France, 1988), p. 579.

historical centres or other historical settings, instead of just isolated monuments, broadened this cultural dimension.

In the 1980s and 90s the 'urban project' arose in parallel to 'strategic planning', 'urban marketing' and the active aggressive action of the local governments in partnerships with private agents. In the urban projects of concentrated isolated intervention, large resources are invested in some structures or buildings endowed with mediatic visibility, which are considered to be capable of spreading 'positive contaminations' to the surroundings or of contributing to the establishment of a new urban image. For Portas[9], who follows up on Venuti, these would be the urban third generation projects, with which are associated the terms 'urban requalification' and 'urban regeneration'.

In this association of corporate planning, urban project and cultural strategy with marketing, there arises an important turning: the culturalist approach of the 1960s becomes a 'market culturalism'[10], in which everything related to culture turns into merchandise. In this metamorphosis, culture becomes the big business of the merchandise-city, which in turn becomes increasingly spectacular. Consequently, two turning points must be considered with respect to the articulation of culture to planning and urbanism: the first one involves revitalisation associated with memory, the heritage and local demands, while the second one involves the mercantilisation, globalisation and spectacularisation of the city and culture.

Bianchini[11] shows how the socio-economic and political contexts of these turning points are also reflected in the changes that have occurred in the field of cultural policies. This unimportant, neutral non-politicised issue of the 1950s and 60s underwent a transformation after 1968, when there arose an association of cultural action with political action. The cultural policies of the 1970s, marked by the emphasis on community development, participation, the democratisation of the public space, and the revitalisation of social life through cultural animation and urban redesign, were replaced in the 1980s. In the climate of neo-conservatism and neo-liberalism, the cultural policies ceased to respond to objectives of the social movements, addressing instead objectives of economic development. They did so, however, not just as instruments to diversify the local economic base or to achieve social cohesion. The subsidies gave way to incentives and exemptions for investments, the social movements to partnerships, planning to the urban project, and renewal and revitalisation to urban requalification and regeneration. The latter, with the aim to maximise the local economic potentialities, placed emphasis on the urban image and on emblematic cultural projects (festivals, exhibitions, the annual promotion of the European capitals of culture, outstanding cultural edifices, etc.).

---

9   Nuno Portas, 'L'emergenza del progetto urbano'.

10   O. Arantes, 'Uma estratégia fatal. A cultura nas novas gestões urbanas', in O. Arantes, C Vainer and E. Maricato, *A cidade do pensamento único. Desmanchando consenso* (Petrópolis: Editora Vozes, 2000), p. 48.

11   F. Bianchini and M. Parkinson, *Cultural Policy and Urban Regeneration – the West European Experience* (Manchester: Manchester University Press, 1993).

In the study of these processes, Bianchini[12] identified U.S. precursor influences of interventions in the historical areas and waterfronts of Baltimore, Boston and New York. Even so, the European cases of insertion of high-visibility cultural amenities became paradigmatic, including the Centre Georges Pompidou in Paris, the Contemporary Art Museum and the Contemporary Culture Centre of Barcelona, and the Guggenheim Museum Bilbao. One often observes a blending of the two principles of revitalisation: the recovery of the existing historical setting, and the creation of cultural amenities as project anchors. In the case of preserved historical settings, the selfsame buildings allude to the local culture; in the case of new architectures, it is their use that gives the seal of culture. These anchors are encircled by magnificently designed public spaces where works of public art are installed and cultural animation actions are carried out. Moreover, these diverse elements are articulated, synthesizing themselves into a new image of the city, which is potentiated through marketing.

In the 'revitalised' areas, the creation of cultural, touristic and recreational activities, of amenities such as museums, galleries, theatres, etc., of festivals and of a commercial ambience of the *fun-shopping* type, provides a consumerist setting which Meyer[13] designates as 'culturalised urbanism'. For Häusserman,[14] the fact that culture is used at present as a magical product utilised by urban marketing results in a 'culturisation' of the city. The term 'cultural regeneration' originated in the Anglo-Saxon setting, referring to the interventions in areas consolidated through this modality of urban planning and urban project, in which culture stands out as a leading strategy and the emphasis of the urban policies is placed on cultural policies.

The effect of this 'urban renaissance' is such that the relationship between the urban dynamic and the presence of culture, mainly involving the visibility of cultural amenities and of the historical and cultural heritage, underwent a radical alteration. In the past, rich cities alone had amenities and a cultural heritage demonstrating their high level of cultural development. Today, however, the recourse to the exhibition of this culture is merely a means, an attempt to achieve a supposed development.[15]

Culturalised planning, through cultural regeneration, has been presenting different results and assessments. Apparently, the less ambitious and less mediatic projects have achieved less dissemination and critical attention. The most spectacular projects gave rise to numerous publications of both the advertising and critical type, to such an extent that specific new terms have been observed in various discourses.

Authors of diverse provenance have analysed the results of these interventions, apprehending, designating and criticising new socio-cultural and spatial trends. Despite their diverging outlooks and some differences in conception, we can mention

---

12 Ibid.

13 H. Meyer, *City and Port. Urban Planning as a Cultural Venture in London, Barcelona, New York and Rotterdam: Changing Relations between Public Urban Space and Large-Scale Infrastructure* (Rotterdam: International Books, 1999), p. 44.

14 Häussermann, *Grossstadt. Soziologische Stichworte.*

15 Evans, *Cultural Planning, an Urban Renaissance?*

and briefly sketch those which we consider to form the main trends. Among those observed in historical centres, one finds excesses of heritagisation,[16] of musealisation and museumisation[17] and of disneyfication.[18]

Considering not only the specific cases of revitalisation of historical centres but also those involving the creation of new avant-garde architectures, a few more trends may be listed, despite the risk of over-simplification and excessive decontextualisation. Among those of a socio-spatial nature, we may mention the aforementioned trend of culturalisation,[19] in addition to those of the monumentalisation[20] and the aestheticisation[21] of spaces. The practice of marketing, of advertising the local identity and image, and even the belief that visibility is equivalent to success, and the consequent amplification of this visibility through the media, has come to be called mediatisation.

Two trends, however, appear to be recurrent: gentrification[22] – the expulsion of the existing population due to the revaluation of the properties in an area –, and the spectacularisation of the city (in the sense defined by Debord[23]), which we are all witnessing, stupefied, with a consumerist, alienating and participationless passivity. The proliferation of images, events, festivals, architectural icons, 'revitalised' and redesigned public spaces, whose symbolic dimension is potentiated and ennobled by culture, becomes the raw material for urban marketing. Culture and the revitalised city are advertised, forming a spectacle to be consumed.

---

16   Excessive attribtuion of the status of heritage. See Henri-Pierre Jeudy, *Espelho das cidades* (Rio de Janeiro: Casa da Palavra, 2005), Brasilian translation of *La machinerie patrimoniale* (Paris: Sens and Tonka, 2001) and *Critique de l'esthétique urbaine* (Paris: Sens and Tonka, 2003).

17   Multiplication of museums in historical properties and excessive designation of historical properties, turning the entire city into a museum. See Andreas Huyssen, *Seduzidos pela memória: arquitetura, monumentos, mídia* (Rio de Janeiro: Aeroplano Ed./ Universidade Candido Mendes/ Museu de Arte Moderna, 2000); and Jeudy, *La machinerie patrimoniale*.

18   Creation of an image reminiscent of a theme park. See Sharon Zukin, 'Paisagens urbanas pós-modernas: mapeando cultura e poder', in *Revista do Patrimônio Histórico e Artístico Nacional*, 24 (1996): pp. 205–219; Michael Sorkin (ed.), *Variations on a Theme Park: the New American City and the End of Public Space* (New York: Hill and Wang, 1992); and Huyssen, *Seduzidos pela memória: arquitetura, monumentos, mídia*.

19   Meyer, *City and Port. Urban Planning as a Cultural Venture in London, Barcelona, New York and Rotterdam: Changing Relations between Public Urban Space and Large-Scale Infrastructure*; and Häussermann, *Grossstadt. Soziologische Stichworte*.

20   From the use of the monumental scale. See Jeudy, 2005.

21   From the use of aesthetic effects. See Jeudy, 2005.

22   Elitisation, expulsion of the poorest population, a term developed by Neil Smith, *The New Urban Frontier, Gentrification and the Revanchist City* (London: Routledge, 1996). See also Arantes, 'Uma estratégia fatal. A cultura nas novas gestões urbanas'; and Zukin, *The Culture of Cities*.

23   G. Debord, *A sociedade do Espetáculo* (Rio de Janeiro: Contraponto, 1967).

Urban Spectacularisation

The present times of crisis of the notion of the city become visible mainly through the ideas of the 'non-city': either through freezing – the museum-city and unbridled heritagisation – or through diffusion – the generic city and generalised urbanisation. These two currents of contemporary urban thought, despite their being apparently antagonistic, tend toward a quite similar result, which may be called the spectacle-city or the 'spectacularisation' of cities.

The most conservative current – late post-modernist or neo-culturalist –, radicalises the post-modern concern for pre-existing cultures, and advocates the petrification or pastiching of the urban space, mainly in historical centres, leading to museumisation, heritagisation or the emergence of the theme park-city and of urban disneylandisation, which are typical examples of the spectacle-city.[24] The so-called progressivist or neo-modernist current follows up on some modernist principles, mainly including the idea of the *Tabula Rasa* (without the same social or utopian concern of the first modernists) and makes an apology for the large scale (XL)[25] and for chaotic urban spaces, which are generally peripheral, or of cities on the world's periphery: *junkspaces*, generic cities, shopping cities or terminal spaces of wild capitalism, which are also displayed in a totally spectacular way.[26]

This quasi-schizophrenia of the contemporary discourses on the city often emerges simultaneously in one same city, with preservationist proposals for the historical centres, which turn into tourist receptacles, and with the construction of new districts *ex-nihilo* in the peripheral expansion areas, which become products for real-estate speculation. The actors and sponsors of these proposals are often also the same, just as there is a similarity between the non-participation of the population in their formulations and the resulting gentrification of areas, demonstrating that the two aforementioned antagonistic currents are two sides of the same coin: the spectacular mercantilisation of cities.

Indeed, the mercantilisation of cities is on the rise, growing more pronounced in recent decades with the withdrawal of the State's action and the increasing presence of the market in urban policies. The neo-liberal policies and the public-private partnerships make the urban actions and interventions of corporate character steadily more viable. This mercantilisation becomes spectacular in the contemporary urban policies and projects, mainly within the logic of strategic planning, on prioritising the building of a unique image of the city. This image should be the result of a city's

---

24 Today the most widespread current in this direction is the so-called New Urbanism, with such projects as Celebration City, built by Disney Corporation.

25 An allusion to the neo-modern 'Bible', the book *S,M,L,XL* (New York: The Monacelli Press, 1995), by one of the foremost representatives of this current, the Dutch architect Rem Koolhaas.

26 A good recent example of this type of spectacularisation was the exhibition Mutations (2000/2001) in Bordeaux. See the catalogue published by ACTAR and Arc en Rêve, Barcelona/Bordeaux, 2001.

own culture, of the so-called 'identity' of the city, but paradoxically these images of distinct cities, with distinct cultures, are growing increasingly similar.

This contradiction may be explained: ever more often, these cities are adopting policies that need to follow an international model imposed by the multinational financiers of the great urban revitalisation projects. On replicating, the model becomes extremely homogenising, creating similar urban images and spaces. This model advocates measures adopted in other cities that came to attract capitals, companies, inhabitants and, above all, tourists, advertising them as highly successful interventions in urban marketing terms.

This model addresses basically the international tourist – not the local inhabitant –, and demands a certain world standard, a standardised type of urban space. Just as already happens with the standardised spaces of the big international hotel chains or even of airports, fast food chains, shopping centres, theme parks, gated communities and even the so-called cultural or entertainment amenities, whether they are museum franchises, cultural shopping places or networks of multi-screen cinemas. All this contributes to causing all the great world cities to resemble one another increasingly, as if they all formed a single image: urban landscapes that are identical or perhaps even, as Rem Koolhaas says, generic. In contrast to the inhabitant, the tourist does not appropriate the space – he simply crosses it. In this case, culture as the expression – or better said – simulacrum of a city's 'identity', is used as an urban instrument mainly to attract tourists.

Thus, the memory of the local culture – which was to be preserved in the beginning – is lost, and in its place are created large touristic 'stage sets'. Most often, the local population is expelled from the intervention area by the gentrification process. On the wealthy peripheries, in the new gated communities, this does not come to happen, since these areas are designed within an idea of social segregation, and even offer a total surveillance level on an international standard of security, which also serves to justify a wide-ranging process of privatisation of public spaces, as has been occurring systematically in the majority of the expansion areas of our contemporary cities. We may also mention the new tourist resorts, mainly of the beach cities, where the same logic becomes installed; the spectacular undertaking on the Costa do Sauipe, for example, near Salvador (Bahia State), which follows the model of the 'new urbanism' in the United States, is also a paradigmatic case of privatisation, mercantilisation and spectacularisation of the urban space – in this case, a city formed solely of hotels, shops and services addressed to leisure – exclusively for touristic purposes.

The contemporary process of spectacularisation of cities is inseparable from urban marketing strategies, which are called revitalisation strategies, seeking to build a new image for the city which will guarantee its place in the new geopolitics of the international networks. What is sold today internationally is, above all, the brand image of the city. The competition is fierce and the city councils do their best to sell the brand image, or logotype, of their towns, basically lending priority to marketing and tourism through their greatest lure: the spectacle.

In Aphorism 34 of Guy Debord's classic book *The Society of the Spectacle* from 1967, it is already announced: 'The spectacle is capital to such a degree of accumulation that it becomes an image.' Three phases could be said to mark urban spectacularisation: the initial phase of beautification or modernisation of cities, when the modern urban images start to be moulded; after that they begin to be sold as simulacra – the case of Las Vegas as studied by Venturi is a classic – and what is sold today is more than just the image of the city: it is the city's brand image. In a word, this brand synthesizes an object of desire and connects the consumer to it.

New types of professionals arose to deal with this complex commercial operation: the international urban marketing consultants, whose mission is to differentiate the perception of the product or service (in this case, the city), re-elaborating, resignifying or rebuilding its image. The success of these practices is such that even the institutions that formerly considered them unethical, surrendered to their call: religious institutions and those of higher education and culture, as well as cities, regions and nations.[27] In the case of the creation of the new urban image – the brand image –, culture is used as an instrument of urban spectacularisation, which serves both for real-estate speculation and for political propaganda.

Critical Reflections and Alternatives

Faced with the apparent consensus on the contemporary city, that is, with what may be called the 'city of single thought',[28] a pertinent criticism may perhaps be more urgent, in view of the current scene, than new models or paradigms. The situationist urban thought,[29] and primarily its criticism of urbanism as a discipline of spectacle, could still be seen today as an invitation to reflection. The situationist ideas on the city, mainly against the transformation of urban spaces into stage sets for touristic spectacles, lead to a clear hypothesis: the existence of an inversely proportional relationship between spectacle and popular participation. That is to say, the more spectacular the urban interventions in urban revitalisation processes, the lesser the participation of the population will be in such processes, and vice versa. This is not an absolute equation, however, since variations may also arise in the proportion of spectacularisation: the more passive (less participative) the spectacle is, the more the city becomes a stage set and the citizen a mere extra; and inversely, the more active the participation is, the more the city becomes a stage floor and the citizen a leading actor instead of a mere spectator. Moreover, the relationship between spectacularisation and gentrification would also be directly proportional,

27   James. B. Twitchell, *Branded Nation: the Marketing of Megachurch, College Inc., and Museumworld* (New York, London, Toronto, Sydney: Simon & Schuster, 2004).

28   See Otília Arantes, Carlos Vainer and Ermínia Maricato, *A Cidade do pensamento único* (Rio de Janeiro: Vozes, 2000).

29   See: Internacional Situacionista, Paola Berenstein Jacques (org.), *Apologia da deriva, escritos situacionistas sobre a cidade* (Rio de Janeiro: Casa da Palavra, 2003), and Debord, *A sociedade do Espetáculo*.

since the urban spectacularisation process is always accompanied by a sort of spatial gentrification, with the expulsion of the poorest people from the intervention areas.

The persons excluded from this spectacularisation process possess perhaps the key to its reversal, which would entail, as suggested by the situationists, popular participation itself. Slums, for example, would be the foremost example of such popular participation[30], since the inhabitants are the persons truly responsible for the effective construction there, as opposed to the inhabitants of the formal city, who very rarely feel involved in the building of their urban space or, in particular, of the public spaces of their city. These areas, the slums, would be authentic 'war machines'[31] against urban spectacularisation. In a certain sense, these 'war machines' – which are alternative forms of resistance or fissures in the globalised system – still succeed in evading the spectacularisation process. The Brazilian cities, in general terms, perhaps even owing to their informality, still manage to maintain some type of diversity, of multiplicity in the urban space. Despite their being subject to the homogenising compressive role of the spectacle-city, urban social actors still succeed in reversing the process by appropriating public spaces for dwellings or for diverse gatherings.

The eagerly craved urban (re)vitalisation – bearing in mind that the meaning of revitalisation here is no longer economic but rather social and cultural, considering vitality to be the life resulting from the presence of a public and of diversified activities – can only be achieved in a non-spectacular way when there is a participative popular appropriation of the public space. Evidently, this is something that cannot be completely planned, predetermined or formalised. Consequently, what would some of the alternatives be to the urban spectacle? We have some clues: a participation in, an effective experience of and a deep acquaintance with urban spaces. These alternatives would necessarily involve the selfsame physical experience of the city, which is almost impossible or completely artificial in spectacularised cities. Moreover, they would involve the sensory experience of the city. Only the individual or collective sensory experience that does not let itself be spectacularised, would not let itself be reduced to mere images. The city would not only cease to be a stage set but it would also become a stage floor for the action of its inhabitants.

To explore these clues on alternatives to spectacularisation, we may refer to two brief case studies on Brazilian metropolises: Salvador da Bahia and Rio de Janeiro. In Salvador, the earliest capital of the Portuguese colony and the most representative nucleus of Afro-Brazilian culture, we may consider the case of the preservationist intervention in Pelourinho, a part of the city's historical centre. In Rio de Janeiro, the capital of Brazil for two centuries, considered a synthesis-image or visiting card of

---

30   See Paola Berenstein Jacques, *Estética da ginga, a arquitetura das favelas através da obra de Hélio Oiticica* (Rio de Janeiro: Casa da Palavra/Rioarte, 2001).

31   Cf. Gilles Deleuze and Félix Guattari, *Mille Plateaux* (Paris, ed. Minuit, 1980). Portuguese translation: *Mil Platôs – Capitalismo e Esquizofrenia*, São Paulo, ed. 34, 1995.

the country as a whole,[32] we may take the case of Lapa, a neighbourhood adjacent to the Central Business District. The Pelourinho project involves a vast preservationist intervention that unfolded recently on a substantially degraded area, while the Lapa project involves a long process of small actions and interventions that unfolded over decades.

The Historical Centre of Salvador – Pelourinho –, designated a World Heritage Site by UNESCO, has been undergoing restoration since 1992 and is now in the 7th stage of the Plan, which has the assistance of the BID.[33] This intervention involves a heritagisation process, with the restoration of façades, a change of uses of old mansions, the rehabilitation of old public spaces and the creation of new ones. These changes are accompanied by an intense process of gentrification, with the removal of over 2,000 families to make way for restaurants, bars and souvenir shops for tourists, forming part of a broader agenda, that is, of a strategic plan of the State tourism agency. The plan even has a cultural animation programme – 'Pelourinho Night&Day' – at the squares that have been created (using the old courtyards of the colonial houses), seeking precisely to lend 'vitality' to the area. The new vitality that has been invented for tourists, with fancifully-attired Bahiana women for photos and exhibition rounds of the regional 'capoeira', is totally artificial: the centre was always 'alive' because it was and had always been intensely lived in (even when it became degraded after its abandon by the wealthier population, who moved to the new modern centralities) by its inhabitants, now expelled without ceremony from the area. It may be observed that many of the ex-inhabitants expelled from the centre, return to this area today as pedlars and spend their nights under the awnings of Baixa do Sapateiro, the nearest commercial area. These people are the memory of the place: the local culture is preserved in the body and skin of the inhabitants. The blood of their ancestors, tortured in Pelourinho ('the Pillory'), was also duly cleansed, hygienicised, to make the area more pleasant for tourists.

The project of intervention, called 'revitalisation' or 'restoration',[34] of the Historical Centre of Salvador, by expelling from the area the inhabitants and their respective popular everyday practices and replacing them with touristic cultural simulacra, destroys the local culture and memory themselves.

Here we may refer to another urban revitalisation experience based on culture: that of Lapa, in Rio de Janeiro. Lapa is a centrally-located historical residential area; it is degraded and popular, a place lived by its inhabitants, working people, immigrants, rogues, and transvestites, and a place famous for its long bohemian tradition. After some devastating renewal interventions carried out in the 1970s, the

32 Carlos Lessa, *Rio de todos os Brasis (uma reflexão em busca da auto-estima)* (Rio de Janeiro: Ed. Record, 2000).

33 Banco Interamericano de Desenvolvimento (Interamerican Development Bank) finances economic, social and institutional development projects in Latin America and the Caribbean.

34 See Márcia Sant Anna, 'A recuperação do centro histórico de Salvador: origens, sentidos e resultados', *Revista RUA*, 8/July–December (2003) and Marco Aurélio Gomes (org.), *Pelo Pelô: história, cultura e cidade* (Salvador: Edufba/PPG-AU, 1995).

area was included in the first urban environment preservation plan of the city of Rio, the Cultural Corridor Project, in the 1980s, safeguarding the existence of a large number of historical properties. In that period, several old mansions belonging to the State were made available for cultural activities, and a fragile alternative cultural amenity, called the Circo Voador (Flying Circus), came to host music and dance shows under its degraded canvas, attracting a young population who had long been absent from that area. The city council made its contribution by carrying out works of infrastructure, paving, lighting, etc. and building a new venue for the Circo Voador. The State followed up on earlier projects for the creation of a cultural district in the area. Diverse instances of the public authority supported initiatives for fairs, shows and public art in the open spaces, beside long-preserved historical monuments or along old streets. Over the years, a popular artist of the area used ceramic tiles to decorate a stairway of the neighbourhood, which became more of a place of attraction and agglomeration of persons. Private initiative came to create such establishments as bars, restaurants and show houses, which joined ranks with the traditional cultural institutions such as recreational clubs, antiquarian-bars[35] and the venues of various local performing groups in the fields of theatre, music, dance, circus and others. New people were attracted and in the 2000s the area has become a new hotspot of culture and bohemia in Rio, visited at weekends by tens of thousands of persons who are mainly seeking Brazilian popular music, a festive ambience and outdoor entertainment.

An article[36] in the press recently discussed the rebirth of the area, pointing out one aspect in particular: the diversity of the groups of people frequenting the area, basically formed by 'the poor and the people who think they are not poor'. The article even went so far as to state that 'Lapa is the place where the two sides of Rio meet, where the split city reconverges'. It recalls that when these populations come together, the city flourishes, and when they draw apart, the city becomes degraded. Despite the author's optimism, since the presence of the more popular strata diminishes as improvement progresses, the important thing to note is the revitalisation process, which has unfolded almost without gentrification. This is a rather particular process, in which there converged a large number of actions directed by different social agents – fortunately without major proposals of interventions, investments or marketing – and which succeeded in generating its own socio-spatial dynamic.[37] Lapa, even if partly gentrified, and while drawing a steadily increasing number of tourists, continues to be a lively place and one that is lived and appropriated by the people of Rio.

---

35 Several antiquarians began operating as bars at night. The old furniture on sale during the day accommodated the clients, who came after shop hours to drink, dine and listen to music.

36 E. Gaspari, 'A linda lição da Lapa', *Jornal O GLOBO*, 10/04/2002.

37 L.F. Vaz and C.B. Silveira, *A Lapa boêmia do Rio de Janeiro: um processo de regeneração cultural? Projetos, intervenções e dinâmicas do lugar*. In press.

We believe that to avoid designing spectacularised spaces, urban experts should also experience and appropriate urban spaces. They should relate to or experience physically the city in itself, their object of action. The distancing or detachment between the subject and the object, between professional practice and the physical acquaintance with and experience of the city, proves disastrous on eliminating the most urban characteristic that the city space possesses, that is to say, precisely, its human nature. And this distancing proves even more disastrous on eliminating what is most human in humans: their own bodies. The physical body of citizens and the body of the city itself meet through the appropriation of the urban public spaces. Spectacularised space, on the contrary, limits and reduces this meeting, through gentrification, or it makes such a meeting merely artificial and scenographic. The main issue with respect to urban interventions would not lie in the selfsame requalification of the material physical space – a pure construction of stage sets – but rather in the type of use to which the public space is devoted, that is to say, in the public appropriation of these spaces. It is only through an effective participation that the public space may cease to be a stage set and transform itself into an authentic urban stage floor – a space for exchange, conflict and meeting.

Chapter 14

Culture, Tradition and Modernity in the Latin American City.
Some Recent Experiences

Roberto Segre

Roberto Segre is professor of the Master and Phd Courses of Urbanism at the Faculty of Architecture and Urbanism in the Federal University of Rio de Janeiro (Brazil) and professor at the School of Architecture of ISPJAE, Habana (Cuba). He holds a PhD in Sciences of Art from the School of Arts and Letters in Habana, and a PhD in Urban and Regional Planning from the IPPUR/UFRJ in Rio de Janeiro. He has written more than 30 books on Latin American and Caribbean architecture and urbanism.

The Cultural Meaning of Public Space

It was necessary to reach the end of the twentieth century so that politicians, businessmen and town planners would assume the cultural meaning of the traditional centralities and their importance as an expression of the historic social memory, whose bequest should last for future generations. The balance between community use of public space and the expression of aesthetic values in architectural and urbanistic attributes was one of the basic characteristics of the colonial city, which lasted up until the nineteenth century.[1] If the expansion beyond the walled limits of cities, characterised by avenues and neoclassical parks, prolonged towards the suburb the refined setting of the bourgeoisie recreational life, it would be at the start of the twentieth century that the eclectic monumentality changed the scale of the centrality, adapted to the new rituals established by local elites.[2] The assimilation of European models in the Latin American capitals transformed the introverted Plazas Mayores (Principal Squares)[3] in a monumental exhibition of public buildings,

---

1   Fernando de Terán (org.), *La Ciudad Hispanoamericana. El Sueño de un Orden* (Madrid: CEHOPU, Centro de Estudios Históricos de Obras Públicas y Urbanismo, Ministerio de Obras Públicas y Urbanismo, 1989).

2   Ramón Gutiérrez, *Arquitectura y urbanismo en Iberoamérica* (Madrid: Ediciones Cátedra, 1983).

3   Juan Carlos Pérgolis, *La Plaza. El centro de la ciudad* (Bogotá: Universidad Nacional de Colombia, Editorial Stoa Libris Ediciones, 2002).

representative of the republican power, although articulated with the colonial inheritance. Nevertheless, the language of the classical orders maintained the cultural and aesthetic values of the bourgeoise city up until the 1930s.[4] The exception to this would be the total demolition of the buildings dating from the colonial period in Rio de Janeiro (1922), when the municipal government devastated the *Morro do Castelo* that contained them, in order to create space suitable for offices and governmental buildings in the central area.

The irruption of modernity went hand in hand with questioning the historic inheritance and the proposal of an urban typology based on the "*à redents*" block system and Cartesian tower blocks designed by Le Corbusier, submerged in large open spaces for the recreational and leisure use of the community at large. The centre abandoned its ancestral leisure function identified with Baudelaire's *flanêur*, was transformed into the environment of international financial capital and the exhibitionist environment of the commerce and consumption. Radical interventions that were present in Le Corbusier's Plan Director de Buenos Aires with Juan Kurchan and Jorge Ferrari Hardoy (1938); in Antonio Bonet's Barrio Sur of the Argentinian capital (1956) and in José Luis Sert's Plan Director de La Habana (1953).[5] These proposals – fortunately not carried out – would have swept away the historical context, conserving some isolated colonial monuments, seeking to shape the future projection of the modern city. This example was applied in Caracas, Santiago de Chile, Bogota and Lima, in cities that lost significant areas of their traditional centres.[6]

In spite of the destruction which took place during the Second World War in the central areas of London, Berlin, Rotterdam and Warsaw, that forced the formulation of a new image conceived in the disseminated book *The Heart of the City* – elaborated from the CIAM Congress in Hoddesdon (1951)[7] – and based on the North American model of the high office towers in the configuration of the *business district*, a movement emerged defending the traditional centres of Europe.[8] Under the aegis of UNESCO and ICOMOS, policies to safeguard the historic city were defined and were rapidy adopted by Latin American town planners and progressive institutions.

---

4   Jean-François Lejeune, *Cruauté and Utope. Villes et paysages d'Amérique latine* (Bruselas: CIVA, Centre International pour la ville, l'architecture et le paysage, 2003).

5   Roberto Segre (ed.), *Las estructuras ambientales de América Latina* (México: Siglo XXI Editores, 1977).

6   Roberto Segre (ed.), *América Latina en su arquitectura* (México: UNESCO, Siglo XXI Editores, 1975); and Jorge Enrique Hardroy and Margarita Gutman, *El impacto de la urbanización en los centros históricos de América Latina* (Madrid: Editora Mapfre, 1991).

7   Eric Mumford, *The CIAM Discourse on Urbanism, 1928–1960* (Cambridge, Mass.: MIT Press, 2000).

8   Roberto Segre, (2004) 'América Latina urbana. El colapso de los modelos em la crisis de la modernidad', in *Conference Book. 11th. Conference of the International Planning History Society (IPHS). Planning Models and the Culture of Cities.* Barcelona, 14–17 July, 2004 (Barcelona: Escola Técnica Superior d'Arquitectura del Vallés, Universitat Politécnica de Catalunya), pp. 57–70.

It can not be put down to chance that over the last two decades multiple collections of historical colonial buildings of the region have been declared as World Heritage sites, including La Habana, Ciudad México, Lima, Puebla, Guanajuato, Cartagena de Indias, Cuzco, Olinda, Ouro Preto, Zacatecas, Valparaíso, Cuenca, Quito and Trinidad.[9] To the cultural content of conservation and restoration programmes, the increasing importance of the tourist flow in the region can be added, that constituted an important economic motor to carry out significant architectural and urban development projects, in some cases influenced by the harmful influence of the Disney model.[10]

In this chapter, a number of interventions carried out in some capital cities of the region are selected. This does not signify that other interventions of a similar nature are of lesser importance, amongst which it is possible to cite the restoration of Viejo San Juan in Puerto Rico and the meticulous work carried out by the Historian of the City of Havana, both in the West Indies, and the defense of the extraordinary urbanistic homogeneity of Cartagena de las Indias, in the Colombian Caribbean. In the Andean axis, the vitalisation and refuntionalisation of the centres of México, Bogotá, Lima and Quito, traditionally occupied by the poorest strata of the population and deteriorated by traffic circulation and informal commerce, all stand out. In Argentina, Miguel Ángel Roca's initiative in Cordoba, in the 1980s, was pioneering, in the search of an integration between the colonial inheritance and modern architecture.[11] In Brazil, since 1937, the year in which the Institute of National Historic and Artistic Heritage (IPHAN) (*Instituto do Patrimonio Histórico e Artístico Nacional*) was created, the State has defended emphatically the preservation of colonial cities. Nevertheless, it did not turn out to be easy to maintain the traditional space intact in metropolises, like Rio de Janeiro, Sao Paolo, Porto Alegre and Belo Horizonte, under the pressure of speculative interests and of models of metropolitan modernity. Between the 1980s and 1990s, the municipal governments carried out significant works: the creation of the "Cultural Corredor" in the centre of Rio de Janeiro and the recent operations of morfogenesis of the principal monuments of Sao Paolo which involved the re-use of existing buildings, such as the National Art Gallery (Pinoteca Nacional), the "Julio Prestes" Railway Station which was converted into a concert hall and the Mackenzie building of the electric company which was transformed into a shopping centre.[12] These all represent examples of the shared consciousness concerning the importance and meaning of urban culture as an indispensable framework of everyday life.

---

9   Joseph L. Scarpaci, *Plazas and Barrios. Heritage Tourism and Globalization in the Latin American Centro Histórico* (Tucson: The University of Arizona Press, 2005).

10   Roberto Segre and Antonio Vélez Catraín, 'Por qué hablar de modelo europeo de ciudad en América Latina?', *Revista de Occidente*, 230–231, July–August (2000): pp. 11–24.

11   Miguel Angel Roca, *Obras y textos* (Buenos Aires: Editorial CP67, 1988); and Miguel Angel Roca, *De la ciudad contemporánea a la arquitectura del territorio* (Córdoba: Ediciones Eudecor, 1997).

12   Nestor Goular Reis, *São Paulo. Vila, Cidade, Metrópole* (San Pablo: Prefeitura de São Paulo, 2004).

Figure 14.1   View of the original waterfront warehouses, transformed into offices, restaurants, etc., in Puerto Madero, Buenos Aires, 1995.

Photo: Segre.

The Return of Gardel: Puerto Madero in Buenos Aires

Carlos Gardel and the port constitute two iconic references of Buenos Aires, associated with tango and the name "porteños" pertaining to the inhabitants of the Argentinian capital. At the same time, they establish the umbilical cord with Europe, as much through the maritime crossing to reach the country – Borges affirmed that if the Mexicans descended from the Aztecs and the Peruvians o from the Incas, the Argentians descended from ships –, as in the sad metaphor of nostalgia, contained in the tango "*Volver*" (to return). From that, while the recovery and morphogenesis of the old warehouses of Puerto Madero – that in Le Corbusier's Plan Director were condemned for demolition –, constituted, in spite of the "gentrified" nature of the urban development operation, a positive factor for the reactivation of the public space of the city of Buenos Aires. This initiative took place in the 1990s – during the government of the President Carlos Menem – and was associated with the urban renewal policies carried out in many of the "world" capitals, in search of an economic return through *marketing* and globalised financial capital investment. It constituted the greatest investment of capital in Latin America – approximately two billion dollars, coming from public and private capital sources (Soros, Eurnekían, Chab-Terrab) –, based on the "Barcelona" model;[13] on the English waterfront initiatives developed in Liverpool and on London's *Docklands*; and on the New York initiatives of *Battery Park* and *South Sea Port* in Manhattan.

The recycling of the brick and metallic structured warehouses lying along an axis of 25 street blocks – belonging to the *functional tradition* and similar to those built in the English ports in the nineteenth century, – allowed for the insertion of offices, restaurants, university and cultural installations and costly residential *lofts*. This concentration of differentiated activities attracted the headquarters of transnational businesses, the development of which were carried out by local and foreign architects, pertaining to the profesional *jet set*: César Pelli built the República tower and the *Bank of Boston*; Kohn, Pedersen and Fox with Hampton and Rivoira the Telecom offices; and HOK International and Aisenson, the Malecón building. Luxurious aprtment blocks and 5 star hotels were built on the 175 ha of adjoining vacant land that contained the railway silos and beach, between the warehouses and Río de Plata, including Mario Roberto Álvarez's Hilton and the recent Porteño Building (2004) by the designer Philippe Starck, contained in a recycled silo, in which the rustic exterior brick surfaces contrast with the sophisticated *kitsch* interior. The architectural and scenographic *marketing* was completed with Santiago Calatrava's pedestrian bridge and Rafael Viñoly Fortabat museum. The elitist character of Puerto Madero coincided with Menem's utopian dream, who declared that Argentina belonged ti the First World, associating the local peso to the dollar by way of national legislation, a

---

13   Joan Busquets and Joan Alemany, *Plan Estratégico de Antiguo Puerto Madero* (Barcelona, Buenos Aires: Ayuntamiento de Barcelona, Corporación Antiguo Puerto Madero, 1990); and Jordi Borja and Zaida Muxí, *L'espai públic: ciutat i ciutadania* (Barcelona: Diputación de Barcelona, 2001).

speculative device that as consequence led to the greatest economic and social crisis of Argentinian history in the twentieth century.

Nevertheless, Jorge Moscato, an urbanist who participated in the preparation of the development plan, affirmed that this intervention was positive for the city. In the first place, because it held back the outward movement of businesses and international corporations towards the new suburban financial centres – which had arisen throughout the whole of Latin America –; and at the same time managing to attract and establish a higher income population within the confines of the *city*, avoiding its decadence and asset stripping, creating an income that turned out to be significant for the economic balance of the municipal government. In the second place, because the initiative to reactivate the public space on the southern coast of the city, it reverted the process of privatisation favoured by Menem's government[14] on the north coast, resulting in the establishment of exclusive recreational centres. The objective consisted of settling the old Le Corbusierian dream of integrating the city with the Río de Plata – a unique landscape space, and the green and maritime lung of the capital –, defining the new functions of free time, not only along the coast, but also on the artificial island created by the military dictatorship in the 1970s opposite the Southern Coast; conceived to contain an exclusive financial centre. With the return of the democracy it constituted the "Ecological Reserve", a huge green lung, still rustic and wild.

The rescue of the social and cultural content of the public space on the coastal edge of Buenos Aires, as much in the capital as in the suburban extension of the provincial territory, was one of the main objectives of the municipal governments, in the last decade of the twentieth century. On the promulgation of the "Environmental Urban Plan" (1998), projects of parks, cultural and sporting installations were all implemented, carried out by local architects with a high quality of design.[15] It should be mentioned that Claudio Vekstein was responsible for the design of the "Parque de la Costa" ("Coastal Park"), one kilometre in length, in the Municipality of Vicente López, transforming a "no-one's land", into a beautiful assembly of multifuntional landscape.[16]

---

14   Alicia Novick, 'La costa en proyectos', *Revista de Arquitectura*, Buenos Aires: Sociedad Central de Arquitectos, 202 (2001): pp. 58–81.

15   Amongst others, one can cite Baudizzone, Lestard and Varas; Dujovne and Hirsch; Juan Carlos López; Manteola, Sánchez Gómez, Santos, Solsona; Hampton and Rivoira; Mario Roberto Álvarez and Claudio Vekstein.

16   Claudio Vekstein, 'Luciérnagas sobre el río marron', *SUMMA+*, Buenos Aires, 56, August–October (2002): p. 26; and Claudio Vekstein, 'River Coast Park', *PRAXIS 4, Journal of Writing+Building*, (Cambridge, Mass.: Harvard University, Graduate School of Design) (2002): pp. 70–81.

Figure 14.2 Arqs. Marta Kohen and Rubén Otero. Memorial de los Desaparecidos, Montevideo (2000).

Source: Photo reproduced from the site of the Interdenica Municipal de Montevideo. Press and Communications Service.

Democratic Spaces in Montevideo

Uruguay is a country with a population of approximately three million inhabitants, more than the half of which is concentrated in Montevideo, with an urbanisation index of 90%. This signifies that at the start of the twenty-first century, the concept of a "closed" city was substituted by the more inclusive vision, at the country scale, of an "urbanised territory".[17] This small nation, considered in the 1950s as the "Switzerland of Latin America", had a high educational and cultural level, based on the predominance of the middle class within the social structure. In the 1970s, the military dictatorship brought about an economic crisis that generated a strong migratory movement towards Europe and Australia, and the abrupt descent in the standard of living of the population. With the restoration of democracy in the 1980s, the neoliberal governments promoted the integration of the Uruguay in the globalised and neoliberal financial system, aspiring to become a continental economic centre: the permanent headquarters of Mercosur was situated in Montevideo. The progressive forces, drawn together in the *Frente Amplio* political party, fought against this orientation and in the political struggle managed win the elections of Montevideo City Council – in 1990, Dr. Tabaré Vázquez took control, followed in 1995 by the

---

17 Luciano Álvarez et al., *El Montevideo que viene* (Montevideo: Comisión Financiera de la Rambla Sur, Intendencia Municipal de Montevideo, 1999).

architect Mariano Arana, elected for two terms –, and finally the Presidency of the Republic, with the appointment of Tabaré Vázquez in 2004.

This political path conditioned the urban interventions in Montevideo, characterised by their social content, removed from projects associated with foreign investment or the creation of sophisticated recreational and consumption centres. The plans that had been elaborated in the 1950s, that had sought to demolish the historic core and substitute it with tower blocks of offices, hotels and luxury apartments, were brought to a standstill.[18] With the definition of the orientation of the Plan Estratégico de la Ciudad (City's Strategic Plan) (1993), and integrating the community participation in the "Citizens' Agreement for the Development of Montevideo" municipal programme, the following key strategic issues for immediate action were established: a) to improve the living conditions and the public service infrastructure for the social strata with the least resources; b) to rescue the loss identity and aesthetic quality of the public spaces situated in the suburban areas; c) to articulate with the rural and regional territory the expansion of the "compact" city; d) to increase the density the urban texture; e) to improve the urban settlements along the coastal strip of the Río de La Plata, characterised by their recreational and tourist function; f) to expand the public, and green and open spaces; g) to recover and restore the deteriorated buildings existing within the historic area of the "Ciudad Vieja" ("Old City").[19]

Some of the principal interventions carried out in the centre of the city include the revitalisation and design of the urban furniture in the Avenida 18 de Julio; the new configuration of the Plaza de la Independencia, lying on the threshold of the "Ciudad Vieja" ("Old City") and pre-announced by the gridded paving design on the ground; the liberty sculpture of the Plaza 1° de Mayo, situated opposite the Palacio Legislativo, which frames the free space for political demonstrations. Particular importance was granted to the historic centre and to the harbour infrastructure, with the creation of a pedestrian area around the Customs House and Port Market, an important centre for everyday social life. With the support from the Spanish Junta de Andalucía, through the Conserjería de Obras Públicas y Transportes (2000), an ambitious programme aimed towards the recovery of abandoned houses in the "Ciudad Vieja" ("Old City") and in the popular neighborhood of Réus were developed, integrated in speculative housing groupings that had arisen at the end of the nineteenth century. The plan was executed with the work of labourers, drawn together in cooperatives of mutual assistance, in order to creating popular dwellings.[20]

---

18   Liliana Carmona and Julia Maria Gómez, *Montevideo. Proceso planificador y crecimientos* (Montevideo: Instituto de Historia de la Arquitectura, Facultad de Arquitectura, Universidad de la República, 1999).

19   Mariano Arana, *Escritos. Nada de lo urbano me es ajeno* (Montevideo: Ediciones de la Banda Oriental, 1999).

20   Benjamín Nahoum, *Las Cooperativas de vivienda por ayuda mutual uruguayas. Una historia con quince mil protagonistas* (Sevilla, Montevideo: Intendencia Municipal de Montevideo; Junta de Andalucía, Agencia Española de Cooperación Internacional, 1999), pp. 137–149.

The future perspective of expansion of the city is oriented in a western direction, surrounding the deep bay, until reaching the Cerro fortress, a symbolic monument dating from the colonial period and located in a high position with a marvellous panoramic view. As an indication of urbanistic volition, the emotional and ascetic Memorial de los Desaparecidos (Memorial of the Disappeared) (2000), produced by the architects Martha Kohen and Rubén Hillock, in memory to the victims of the military dictatorship, was erected at the foot of the Cerro.

People in the Streets of Santiago de Chile

To those who had the emotional experience of participatng in the popular euphoria during the short-lived government of Salvador Allende will recall the massive presence of the population in the public spaces of the city. Some of these spaces had been doted in the 1970s by the City Council, with recreational and cultural infrastructures: among these one can recall the O'Higgins Park, inaugurated in 1972. To mark the celebration of the Third World Conference on Development and Free Commerce, organised by the United Nations (UNCTAD III) in Chile, the headquarters was built in the centre of Santiago. The "Gabriela Mistral" grouping, configured by an office block with an elaborate basement that contained the cultural activities and

Figure 14.3   Arqs. Undurraga and Deves, Plaza de la Constitución in front of the Palacio de La Moneda. Santiago de Chile, 1985.

Photo: Segre.

the conference rooms became an urban icon of the socialist government. With the end of this international event, it was transformed into a popular meeting place for young people and students, participants in the exhibitions, concerts, stage and film shows.[21] With the destruction of the Palacio de La Moneda under the military coup led by August Pinochet on 11 September 1973 – with the assassination of Allende – the dictatorship was installed in this grouping that was designated Diego Portales (creator of the repressive Chilean police system), and isolated it with a high wall, closing it to all types of public activity.

During the period of the dictatorship social life was restricted in the central area of the city, shifting to the neighbourhoods of the middle-class elite: Ñuñoa, Providencia, El Golf. In the 1970s and 1980s luxurious steel and cristal office buildings arose, occupied by transnational companies that adopted the international models of the globalised financial system, as well as *malls* and *shopping centres* identified by the presence of the sophisticated international and designer *brands* (*grifes*); in the Providencia, Pedro de Valdivia and Vitacura Avenues, one can cite the Santa Maria Tower (1977); San Ramón (1988); the Shell Chile Tower (1989). This process continued during the 1990s, with the El Bosque Tower (1994); the De Capitales Building (1995) and the World Trade Centre (1995). A few of these were carried out by local architects of prestige – Borja Huidobro, J. Echenique, E. Browne, J. Sabbag –, who were interested in any public space with wide footpaths, parks, plazas and underground parking. Amongst the more recent developments the introduction of works of art in the Vitacura Avenue – by Jorge Figueroa and Paola Durruty (2001) – stand out; and the elaborate landscape design of the Peru Plaza in the Las Condes de Sami de Mizrahi Dinar (2002) neighbourhood.

With the end of the dictatorship, the centre of Santiago recovered its vitality, identified with the crowded presence of pedestrians in the streets and squares, and the expansion of informal commerce. In a restricted of the historic centre, the circulation of traffic remained restricted and a hierarchy was granted to pedestrian space through the design of urban street furniture. The restoration of colonial monuments and academic buildings dating from the 19th and 20th Centuries was accompanied by the revaluation of public spaces with a symbolic meaning, along two principal axes. The first connects the Plaza de la Constitución, opposite the Palacio de La Moneda, with the Plaza de Armas, encircled by the buildings of the Cathedral, the Post Office and the City Council. The second, perpendicular, links this nucleus with the cultural center of Mapocho Railway Station. With the restoration of the Plaza de la Moneda, the square was redesigned, and is usually utilised as parking for the official cars. The architects Undurraga and Devés established a diagonal circulation, that allowed for the dynamic perception of the "neoclassical" space, contrasting with the symmetrical monumentality of the palace. The architects Rodrigo Pérez de Arce, Sebastián Bianchi, Álvaro Salas and Leonor Caamaño, devised a refined landscape solution for the Plaza de Armas. This was divided into functional spaces

---

21   Roberto Segre and Rafael López Rangel, *Architettura e territorio nell'America Latina* (Milán: Electa Editrice, 1982).

Figure 14.4   The "Transmilenio" system with the favourite stop in Bogotá. Arqs. Javier Vera, Fernando León Toro Vallejo, Gabriel Jaime Giraldo Giraldo, 2000.

Photo: Segre.

with a distribution of the green areas that facilitated the perceptive appreciation of the monuments and a homogeneous integration between the floor of the square, the footpaths and the streets all at the one level, differentiating between these by the textures of the horizontal plan. Finally, the grouping of the Central Market and the Mapocho station defined a gastronomic and cultural nucleus, characterised by the intense popular participation in organised events in the large open space of the railway station.[22]

Bogotá: Aesthetics and Solidarity

In a country characterised by the continuous violence generated by a prolonged civil war for more than half century, Bogota, with more than six million inhabitants, was

---

22   Humberto Eliash (ed.) *XII Bienal de Arquitectura de Chile. Arquitectura de uso público. Reinventar el futuro* (Santiago de Chile: Colegio de Arquitectos de Chile, 2000).

nevertheless able to witness the achievement of ambitious urban plans and a high aesthetic quality of its architecture. Its history did not turn out to be easy, being marked by a number of tragic events. In 1948 a popular revolt took place – the "Bogotazo" – through the murder of the political leader Jorge Eliecer Gaitán, which destroyed part of the central area. In 1985, the army bombarded the building of the Supreme Court of Justice, situated in the plaza Simón Bolívar, annihilating judges and guerrillas of the M-19 movement, who had occupied the building and who had taken the magistrates hostage. The academic architecture of the original headquarters was substituted by a banal post-modernism, which altered the architectural homogeneity of the city's main square city.

The *campus* of the Ciudad Universitaria (1936), projected by the German architect Leopoldo Rother, constituted the starting point of Colombian modern architecture – based on the language of the "white boxes" –, substituted in the decade of the 1950s by a regionalist expression, based on the generalised use of brick, developed by the architects Carlos Martínez and Rogelio Salmona.[23] In those years, the urbanistic ideas contained in the development plan elaborated by Le Corbusier, Paul Lester Wiener y José Luis Sert were applied in part, in which the application of the road system dominated. Nevertheless, up until the 1990s, the dynamics of the urban growth were defined by technocratic planning, favouring the organisation of the higher socio-economic determined areas based upon the metropolitan models – applied in the office blocks of the National Administrative Centre and of the International Centre,[24] – building anonymous housing estates in the suburbs; without controlling the disorderly public transportation and the growth of the spontaneous peripheral settlements of the lower income social strata.

This somber panorama changed in the 1990s, with initiatives promoted during the term of the then mayor, Enrique Peñalosa Londoño, insisted on improving the public urban space and the aesthetic qualification of the city, objectives that have fortunately been maintained until today (2005) by his successors in the local administration. With the creation of the Capital District in the new Constitution (1991) and the elaboration of the Plan de Ordenamiento Territorial 2001–2010, the organisation of five new local centralities was proposed and the elaboration new urban projects with the following objectives: a) to prioritise the organisation of public transport; b) to install technical infrastructure in the suburban areas and in lower income neighbourhoods; c) to create cultural facilities and in particular neighbourhood libraries; d) to establish a green system at district and neighborhood scales, reintegrating the pedestrian to urban life; and e) to revitalise the traditional centre, abandoned by the outward movement of public functions.[25] The developments, which took place at different scales, were

---

23   Eduardo Samper Martínez, *Arquitectura moderna en Colombia. Época de oro* (Bogotá: Diego Samper Ediciones, 2000).

24   Alberto Saldarriaga Roa, *Bogotá siglo XX. Urbanismo, arquitectura y vida urbana* (Bogotá: Alcaldía Mayor de Bogotá, 2000).

25   María Carolina Barco de Botero (ed.), *Taller del Espacio Público. Proyectos 1998–2000* (Bogotá: Alcaldía Mayor de Bogotá, 2000).

undertaken by professionals of prestige, who guaranteed the aesthetic quality of the architectural works.

The outdated and disorderly system of public transport, based on obsolete and old buses, and privately operated vans, was transformed by the adoption of the Brazilian model of Curitiba, through the creation of a system of road axes giving priority to the circulation of buses with fixed stops. The structure of the Transmilenio, through the functional efficiency, the wise design of the equipment and of the transparent bus stops – designed by Javier Vera, Fernando León Toro Vallejo and Gabriel Jaime Giraldo Giraldo –, changed the image of Bogota's public space; it facilitated the rapid access to the centre of the population, settled in the distant periphery and managed to diminish the presence of the private car. The creation of a system of parks and squares – approximately 900 such parks and squares of different sizes, named as "Parks for learning to live in", were designed and recovered – comprising parks of a metropolitan scale such as Simón Bolívar, El Tunal and San Christóbal; the creation of green axes with cycleways including Alameda Santa Fé, measuring some 80 Km. in length and extending through various neighbourhoods; Alameda El Porvenir measuring 17 Km, designed by Felipe González, Pacheco Mejía, Juan Ignacio Muñoz Tamayo; the establishment of green public spaces in depressed urban areas, such as La Aurora II of Giancarlo Mazzanti, Rafael Esguerra Cleves, Carlos Hernández Correa;[26] and hard surfaced squares in the centre of the city, including San Victorino square elaborated by Lorenzo Fort Jaramillo's "Taller del Espacio Público" and the urban furniture of the Avenida Jiménez de Quesada, of Rogelio Salmona and Luis Kopec.

Finally, the creation of a network of public libraries – with some of them being situated in popular neighbourhoods – turned out to be a successful initiative on the basis of the number of people making use of them, particularly of youths and adolescents. The principal ones were projected by prestigious local architects, amongst which figure Rogelio Salmona's Virgilio Barco; El Tunal by Manuel Antonio Guerrero, Suely Vargas and Marcia W. De Vargas; and El Tintal by Daniel Bermúdez. The experience developed in Bogota constitutes a significant model in Latin America, not only for the aesthetic quality of the projects carried out, but rather principally for the concern to multiply the public spaces of social contact, as much in the higher socio-economic areas as in the poor distant suburbs, within the precarious conditions of equilibrium existing in the neoliberal political system, and contrasting with the spatial and social segregation of the city, imposed by the speculation, the interests of financial capital and the local elites.

Caracas: A City in Conflict

The city is the mirror of the economic, social, political and cultural life of the community that inhabits it. The changes in the physical and spatial structures of

---

26   Rodolfo Ulloa Vergara (ed.), *XIX Bienal Colombiana de Arquitectura* (Bogotá: Sociedad Colombiana de Arquitectos, 2004).

Figure 14.5   Caracas. Partial view of the Avenida Bolívar with the pedestrian porticos opposite the Parque Central. Intervention of Carlos Gómez de Llarena en the 1990s.
Photo: Segre.

cities throughout history coincide with changes in their value and meaning resulting from transformations of social life. Therefore, Caracas is the reflection of the deep tensions and contradictions that have characterised the history of Venezuela. Throughout almost three centuries, the low development of the colonial economy maintained orderly relations with Spain, up until the end of the nineteenth century. In the first decades of the Republic, in the period around the change of the century, the academic character of the public buildings – the Capitol and the University – is associated with a democratic conception of public space, whose climax is reached in the central Plaza Bolívar. In the area of the historic centre, Carlos Raúl Villanueva projected a modern group of housing – El Silencio (1942) –, which maintains the equilibrium between the new architectural forms and the popular use of galleries, streets and squares.[27] With the discovery of petroleum, in the 1920s, the country began to subsist from the exploitation of this natural resource, abandoning the agricultural and livestock exploitation. The incipient industry was revived in the 1950s with the exploitation of iron mines in the Orinoco. The infinite resources available allowed for the design of high speed viaducts and gigantic towers – Siso and Shaw's Parque Central complex (1979) –; and sought to eliminate the precarious spontaneous settlements in the valle de Caracas, substituted by infinite blocks of flats

27   Roberto Segre, 'Múltiplas vozes em Barcelona', *Projeto-Design*, San Pablo, 296, October (2004): pp. 110–113.

(1952), designed by Carlos Raúl Villanueva.[28] The "dual" city, comprising the rich "formal" neighbourhoods of the eastern zone and the poor settlements of the western zone – with more than half the population of four million inhabiting in the hills –, defined a segregated and inhabitable landscape, dominated by the invading presence of the automobile.

In the 1970s and 1980s, a number of diverse urban development initiatives arose that sought to recover the urban "humanity" urban and the loss of spatial and social integration. The creation of the Metro along the valley and with ramifications towards the popular neighbourhoods, allowed for the defunct public transport to be improved and to integrate the different social groups that utilised it. Its high design quality – carried out by a team led by Max Pedemonte –, and the insertion of works of art in all the stations, constituted an important factor in the aesthetic education of the population. At the same time, these constituted elements of urban revitalisation, on creating within their vicinity pedestrian areas and squares in the different neighbourhoods. Another initiative, developed by Carlos Gómez de Llarena, had as its objective the rescuing of the Avenue Bolívar – high speed road axis – for pedestrians; the suspending the construction of office blocks and completing the free spaces with arcaded galleries, squares, and recreational and cultural centres. The creation of the Parque Vargas, and the redesign and refuncionalisation of the office blocks of the Centro Simón Bolívar, being converted into the Palacio de Justicia, created a "system" of functional and articulated spaces, which culminated in the central and popular area of the city. At same time, in the spontaneous settlement of San Augustín, near to Central Park,an attempt was made to organise a centre of services – La Franja de Manuel Delgado (1989) – the design of which was not assimilated by the community. They constituted interventions oriented towards the creation of quality public spaces, which sought to reduce the spatial and social segregation in the urban context.[29]

Some architectural critics – Segre, Gómez, Posani, Niño, Hernández de Lasala – supported these hypothetically successful initiatives. Nevertheless, the urbanist Martin Frechilla (1991) had a more realistic perception through understanding the seriousness of the existing social conflicts, the tensions os which were on the verge of exploding and which invalidated the urban design proposals. This thesis was ratified by the violence which unfolded in Caracas in 1989 (El Caracazo) against the economic measures imposed by the president Carlos Andrés Pérez. The poor inhabitants of the hills descended to the "formal" city, and vandalised stores and shopping malls pertaining to the middle class.[30] From that moment a

---

28 Roberto Segre, 'Carlos Raúl Villanueva. A cidade como obra de arte', *AU, Arquitetura and Urbanismo*, 19/128, November (2004): pp. 63–68.

29 Roberto Segre, *América Latina Fim do Milênio. Raízes e Perspectivas da Sua Arquitetura* (San Pablo: Studio Nobel, 1991).

30 Maria Pilar García-Guadilla, 'Territorialización de los conflictos socio-políticos en una ciudad sitiada:*guetos* y feudos en Caracas', in Ciudad y Territorio. Estudios Territoriales (CyTET) (Madrid: Ministerio de Fomento), XXXV/136–137 (2003): pp. 421–440.

political, economic and social crisis arose in the country, that culminated with the election of Hugo Chávez as the president (1998). Assuming the representation of the less economically favoured classes of the society, Chávez initiated a process of transforming the laws then in force under under the Constitution, establishing "participatory democracy" (1999), the objectives of which were not fulfilled before the violent reaction of the Government's opposition. Social antagonisms ensued that manifested themselves aggressively in the urban space, transforming Caracas in a divided city.

The middle classes, enclosed in their privatised spaces in the eastern zone of the city,[31] converted in feuds and *guettos* of the municipalities identified with the "opposition" – Chacao, Baruta, El Hatillo and the urbanisations of La Florida, El Paraíso, Bellomonte, – created an environment for the development of their political actions, concentrated on the plaza Altamira, designated "plaza de la Libertad", that then extended along the motorways towards the Avenida Bolívar. The "chavistas" occupied the extensive territory of the Libertador municipality in the western zone, arriving at the center and the Avenida Bolívar, converted in a field of confrontation between the two antagonistic groups. On the unsuccessful end to the prolonged strike of 2003, and Chávez's achievement of a significant victory in the "referendum" (2004), the social tensions and the day to day violence were reduced, but the Government's initiatives destined to intervene in the areas of the consolidated city were held up. Without doubt, the political interests are directed towards the improvement of the living conditions of the lower income population, operating in the numerous settlements scattered in the hills. When the "participatory" democracy eventually reaches a state of equilibrium, the city will return to contain the meeting places and places of encounter for the differentiated members of the community.

---

31   Armando Montilla, 'Tomando las calles: Caracas 2002–2003', in *aula No. 4. Arquitectura y Urbanismo en las Américas* (New Orleáns: School of Architecture, Tulane University, 2004), pp. 44–53.

Chapter 15

'Cities Are Fun!': Inventing and Spreading the Baltimore Model of Cultural Urbanism

Stephen V. Ward

Stephen V. Ward is Professor of Planning History at Oxford Brookes University (England). He is the present Editor of *Planning Perspectives* and a former President of the International Planning History Society. He is the author or editor of several books and numerous articles and book chapters on historical aspects of planning. His most recent books are *Selling Places: The Marketing and Promotion of Towns and Cities 1850–2000* (Spon, 1998), *Planning the Twentieth-Century City: The Advanced Capitalist World* (Wiley, 2002) and *Planning and Urban Change* (2nd edition, Sage, 2004).

In 2004 Baltimore recorded, at 41.8 homicides per 100,000 inhabitants, the worst murder rate of any American city.[1] Moreover, this was merely the most shocking headline from the Maryland city's unenviable portfolio of indicators of economic distress, social deprivation and community breakdown.[2] Between the censuses of 1950 and 2000 the city's population fell from just under 950,000 to just over 650,000. More to the point, and unlike many other American cities, this trend of decline has actually accelerated in recent years. Thus the city recorded an 11.5 per cent loss in the 1990s, the worst rate amongst the largest American cities, and not far short of twice its own rate of decline in the previous decade. These same trends were reproduced in other characteristics – low educational attainment, family breakdown, housing abandonment, drug abuse and much else. In 1999, nearly a quarter of its population (22.9 per cent) fell below the poverty line. Behind all this (though certainly not the sole causal factor) lay a huge and continuing loss of jobs. This was overwhelmingly from the manufacturing and related industrial sectors which recorded a decline of

---

1   Suzanne Goldenberg, 'Our youth are killing each other. Every other day it seems someone is murdered', *The Guardian* February 22, 2005.

2   James R. Cohen, 'Abandoned Housing: Exploring Lessons from Baltimore', *Housing Policy Debate*, 12/3 (2001): pp. 415–448; *Baltimore in Focus: A Profile from the US Census 2000* (Washington DC: The Brookings Institution Center on Urban and Metropolitan Policy, 2003).

41.7 per cent between 1990 and 2002.[3] The performances of many types of services were similar with finance recording a 36.3 per cent job decline and retailing a fall of 47.3 per cent over the same period. There has been only very modest employment growth in other services to offset this grim picture.

At this point the reader may quite reasonably wonder how Baltimore could ever possibly have been considered a 'model' that other cities might have wished to emulate. The answer lies in that part of the city that visitors, who still come in their millions, tend to see, namely the Inner Harbor and adjoining blocks. A quarter of a century ago, the extraordinary transformation that was then underway in this area seemed to offer a panacea for the problems of all declining industrial, especially port, cities. It seemed also to offer a vision of an urban economy that could replace the old jobs with a service economy that was based increasingly on leisure and tourism. Manufacturing production and ancillary services were to be replaced by cultural consumption. The grimy workaday cities of the industrial era would be replaced by cities that could be, in the words of Baltimore's greatest promoter, 'fun'. Moreover it seemed to be a way of harnessing private capital to public policy, so that the model could be realised without excessive reliance on public spending. It therefore represented an urban expression of a major theme of global political economy during the last decades of the twentieth century.

For a time, at least, the Baltimore model proved very compelling and was widely received in a very uncritical manner. First other American cities and then many in other countries sought to copy or at least capture something of what their leaders thought they saw in the Maryland city. They were aided by the active promotional efforts of those who were central to the Baltimore experience. This chapter documents the formation of the model and examines its nature. It also sketches its impact on thinking about the transformation of the redundant physical and economic spaces of port and industrial cities throughout the world. In a more specific sense, it also shows how those associated with the invention of the model played a direct role in spreading this model. We also note how far the Baltimore model has become synthesised into the reservoir of global planning ideas and practice. Finally we return to the issues with which the chapter began, considering the extent to which the model was disconnected from the real city whose name it took.

How Did the Baltimore Model Appear?

The story of Baltimore's impact on international thinking about the future city began as early as 1954. At that time there was growing concern within the city's business community about the decline in its property tax base as middle income population and the downtown retail activities their spending underpinned began to leave for

---

3   Cohen, p. 418; Bay Area Economics, *Industrial Land Use Analysis: City of Baltimore Maryland*, Report prepared for Baltimore Development Corporation (Washington: BAE, 2004), pp. 9-13.

the suburbs.[4] The following year this concern led to the formation of a key body for mobilising opinion, the Greater Baltimore Committee (GBC). One of the GBC's most urgent concerns was to renew the downtown area to reflect the changing realities of the wider region. In 1958 it produced its own largely private-financed plan to renew a section of downtown. The result was the Charles Center, an acclaimed mixed-use development. By 1964, the GBC was sufficently emboldened to launch another plan to renew the moribund Inner Harbor which lay nearby. A partial clean up of the deteriorated piers and sheds had begun in 1950, followed by some landfill and road provision, but the long term use remained undetermined. Again produced by the Wallace company, the 1964 plan extended the Charles Center, also proposing new public administration buildings, educational and cultural facilities and housing, together with several visitor attractions.

Compared to the Charles Center, the Inner Harbor plan required much more public investment, especially from federal but also from city sources.[5] Yet the urban political regime within which this public spending occurred continued to be based on public-private partnership. This was enshrined in 1965 when the management of the transformation of the Inner Harbor's transformation was assigned to the same private non-profit corporation that had overseen the Charles Center project whose remit was now extended.[6] The Charles Center-Inner Harbor Management Inc brought together city government, Baltimore's business leadership and the developers of the individual projects. It thus harnessed private expertise and networks even though the developers were in many cases the city or other public agencies.

But this is to anticipate later events. The original 1964 plan did not meet with the widespread approval that had greeted the earlier effort for the Charles Center.[7] The greater call on public funds cast it into the very centre of political debate at a time of growing tension in the city culminating in the race riots of 1968. There was also greater than anticipated resistance to the relocation of public administration offices. Problems were experienced securing federal funding for the middle income housing that was intended to combat the flight to the suburbs. The outcome was that much greater emphasis than originally planned was placed on visitor attractions. This pragmatic and, at times, somewhat contested shift towards a more (but not totally) touristic and cultural form of urbanism was to be crucial in establishing Baltimore's reputation.

Little happened until the election of Mayor Donald Schaefer in 1971. The first tangible sign of the Inner Harbor's changing character was the appearance the

---

4   Katherine Lyall, 'A Bicycle Built-for-Two: Public-Private Partnership in Baltimore', in R. Scott Fosler and Renee A. Berger (eds), *Public-Private Partnership in American Cities: Seven Case Studies* (Lexington MA: Lexington Books, 1982), pp. 17–57.

5   David Wallace, *Urban Planning/My Way* (Washington: American Association of Planners, 2004).

6   Douglas M. Wrenn with John A Casazza and J. Eric Smart, *Urban Waterfront Development* (Washington DC: The Urban Land Institute, 1983), pp. 147–148.

7   Timothy Barnekov, Robin Boyle and Daniel Rich, *Privatism and Urban Policy in Britain and the United States* (New York: Oxford University Press, 1989), pp. 92–95.

Figure 15.1   Baltimore Inner Harbor

following year of the historic warship *USS Constellation* as an attraction. Public access to the waterfront was also created and the city began aggressively promoting an active programme of largely free entertainment. This generation of activity paid dividends by encouraging funders that investments to tap this potential leisure market would be worthwhile. From the mid–1970s, major developments began to appear in quick succession.[8] The Maryland Science Center opened in 1976, the World Trade Center office tower in the following year and the Marina in 1978, all broadly as envisaged in the 1964 plan. Then followed a series of developments which had not been in the plan – the Convention Center in 1979, the Harborplace festival marketplace in 1980 and a new Hyatt hotel and the National Aquarium in 1981. Further development continued in the eastern part of the Inner Harbor from the mid–1980s with a concert hall, museums and other cultural facilities. But by then Baltimore Inner Harbor's reputation as an national and international model of urban regeneration was already established.

   In 1983 alone the city was visited by an estimated 4,000 representatives from 87 cities around the world, eager to learn how its apparent renaissance had occurred.[9]

   8   Wrenn with Casazza and Smart, *Urban Waterfront Development*, pp. 146–155; Ann Breen and Dick Rigby, *Waterfronts: Cities Reclaim Their Edge* (New York: McGraw-Hill, 1994), pp. 113–114.
   9   Joshua Olsen, *Better Places, Better Lives: A Biography of James Rouse* (Washington DC: The Urban Land Institute, 2003), p. 293.

The Appeal of the Baltimore Model

What was it then which drew these official visitors? In some respects, and in common with all urban models, Baltimore's appeal rested in part on a series of illusions. Thus several of its admired features were, variously, not new, not particularly unique to Baltimore or not even what they appeared to be. In public-private partnership, for example, Baltimore was merely following the lead given by Pittsburgh's business and civic leaders in 1943.[10] Even more paradoxically, these aspects had actually been greatly modified by the time Baltimore became an object of international admiration. In 1979 the GBC's capacity for decisive action was weakened when its membership was widened to all business interests.[11]

Even before this time, the financial nature of the partnership was not what it seemed. The harnessing of private expertise, information networks and good will was real enough. But the amount of private *capital* that was actually being risked in the developments was less than appeared (another feature that was not peculiar to Baltimore).[12] In part, this tendency to underplay the public sector role was because of the obscure nature of many public subsidies and tax breaks. It was compounded by the huge attention which settled on a few very high profile private developments that were to play a key part in defining the Baltimore model. These features helped to foster some dangerous illusions for the leaders of many other American cities as public funds were curtailed in the 1980s.

The most significant private development in Baltimore was the Inner Harbor's major single draw, the Harborplace festival marketplace. It was developed by the well known and widely respected Baltimore developer, James Rouse. Paradoxically, most of his activities as a developer were devoted to promoting the movement of people and retailing from cities to air-conditioned suburban shopping malls (of which he was a pioneer) and to suburban residential communities, especially his celebrated private new town of Columbia, Maryland.[13] Towards the end of his main working life, however, he became interested in downtown revitalisation. He was to be central to the creation of the Baltimore model.

In 1972, Rouse teamed up with a Boston architect, Ben Thompson, over the latter's plans to revive the historic but derelict Faneuil Hall-Quincy Market area of Boston. The area lay between the downtown offices and a waterfront that was itself being revived (with a newly opened Aquarium that was itself to be the model for Baltimore's). Thompson's plan involved maintaining the historic fabric and returning it to something closer to the idea of a market rather than following the

10 Shelby Stewman and Joel A. Tarr, 'Four Decades of Public-Private Partnerships in Pittsburgh', in R. Scott Fosler and Renee A. Berger (eds), *Public-Private Partnership in American Cities: Seven Case Studies* (Lexington MA: Lexington Books, 1982), pp. 59–127.

11 Lyall, p. 31.

12 Bernard J. Frieden and Lynne B. Sagalyn, *Downtown, Inc. – How America Rebuilds Cities* (Cambridge MA: MIT Press, 1989), especially pp. 155–170.

13 Nicholas Dagen Bloom, *Merchant of Illusion: James Rouse, America's Salesman of the Businessman's Utopia* (Columbus: Ohio State University Press, 2004), pp. 27–149.

usual shopping mall template. Shops and market stalls were to be small, individual and distinctive, with neither department or chain stores. With some difficulty Rouse won the city's approval and with even greater difficulty he secured financial backing. Actually it was not entirely without precedent because San Francisco's Ghirardelli Square which had some similar features had opened in 1964. But this had been in a much less prominent location and, at the time, had attracted little interest. This contrasted with the much-trumpeted opening and rapid success of the first section of the revived Quincy Market in 1976, with phases two and three in successive years. It triggered a huge interest in the concept which Rouse had by then dubbed the festival marketplace.

One of the first cities to take note was Baltimore, yet again borrowing the innovations of other cities.[14] The head of the Charles Center-Inner Harbor Management Inc was an old acquaintance of Rouse, Martin Millspaugh. He invited the developer to consider a similar development there, again reusing an old building. Though Rouse rejected the building, he pursued the idea on a better located site that took over what had been scheduled as a public open space where the Inner Harbor area met downtown. The loss of the open space was very unpopular and it was only after a closely fought city referendum that Rouse was allowed to proceed. In the first full year of operation Harborplace attracted an astonishing 18 million visitors, with sales per square foot many times greater than in suburban malls. It also exceeded even the tremendous success of its slightly older sister in Boston.

More than any other development, it was Harborplace which sealed Baltimore's reputation as a model. Yet the model was always more than Harborplace. It would be more accurate to say that the rapid completion of several key and individually successful visitor attractions in close proximity both to each other, to what was now an attractive waterfront and to downtown was mainly responsible for Baltimore's reputation. And the really astonishing fact was that all this had occurred within such an unpromising city. Boston's historic role in the formation of the United States meant that it was already an established tourist destination. Baltimore had no such prior bridgehead for economic transformation. Scarcely anyone had previously chosen to go there for any but the most utilitarian of reasons. Therefore its sudden emergence as a leisure and tourist destination in the early 1980s seemed to offer a way in which all industrial port cities could negotiate their way from the grim realities of dock closures and deindustrialisation to the sunlit uplands of a successful post-industrial future.

Like all models, objective reality was also strengthened by a large dose of conscious promotional effort. Rouse's prominence and gift for promoting his developments focused first the nation's and then the world's attention on Baltimore. Flush with this huge commercial success and popular acclaim, Rouse himself was featured on the cover of the prestigious *Time* magazine in 1981.[15] Above his own image and representations of his latest successes was his own slogan: 'Cities are

---

14   Olsen, pp. 278–291.
15   *Time*, August 24th 1981, cover image by Wilson McLean.

fun!' In immediately accessible language, Rouse had captured the essential feature of what is, more cerebrally, here called cultural urbanism. An increasingly leisured society with more time and wealth at its disposal now expected cities to offer more leisure and cultural diversion. And the consumption that they generated could be captured to create a new economic base for deindustrialising cities. The resultant new attractions also recycled the redundant spaces of the industrial era.

In a broader sense Rouse realised that many Americans yearned for a romanticised notion of a vibrant traditional city. In reality this was a place that had never been, a scene of happy animation where people might safely gather in large numbers. It was a scene untroubled by all the competing and troubling realities of the industrial past or the post-industrial present. Here was a carefully managed enclave from which all the many problems of urban decay, crime, social and racial tension had been banished. Here too were cultural entertainment and sanitised fragments of traditional urbanism – the market hall, the handcrafted products and local traders of cities before large scale capitalism. It was no accident that Rouse was a great admirer of Walt Disney and that the two men shared a similar vision of what an ideal American community should be.[16]

The 'Rousification' of the United States[17]

As Harborplace opened, Rouse, the Baltimore model's greatest promoter, actually retired from the company which bore his name and later severed all ties. Yet both he and the company continued to be central to efforts to spread the model Cities wanting to 'do a Baltimore' naturally sought out the Rouse company that claimed such a central role in developing and promoting the concept. The Rouse company was already working on a waterfront development with a festival marketplace at the South Street Seaport in New York.[18] During the early 1980s it took on several other projects, essentially varieties of the festival marketplaces concept, in Miami (Bayside), Atlanta (Underground Atlanta), New Orleans (Riverwalk) and Portland, Oregon (Pioneer Place).[19] It also undertook a further mixed-use development, albeit of more conventional type, at the Baltimore Inner Harbor.

Rouse himself, meanwhile, pursued philanthropic activity through his new Enterprise Foundation established in 1981.[20] To provide an income stream for these this, however, he also established the Enterprise Development Company (EDC) to undertake for profit development activities. Fronting the company alongside Rouse was Martin Millspaugh, along with a few other key figures from Rouse's earlier

---

16   Olsen, pp. 240–241.

17   Peter Hall, *Cities of Tomorrow: An Intellectual History of Urban Planning and Design in the Twentieth Century* (Oxford: Basil Blackwell, 1988), pp. 347–351.

18   John T. Metzger, The failed promise of a festival marketplace: South Street Seaport in lower Manhattan, *Planning Perspectives*, 16/1 (2001): pp. 25–46.

19   Metzger, p. 4; www.premiermarketplaces.com (accessed 23.06.05).

20   Olsen, pp. 298–309.

projects. Millspaugh's presence underlined the EDC's direct association with the Baltimore model in general rather than just Rouse's particular contribution to it. Even so, the reproduction of festival marketplaces formed EDC's major development activity in the early years. However, it also built up a significant business in giving more general advice in broader development aspects of the wider areas in which these marketplaces were set. (It was later renamed Enterprise Real Estate Services.)

There were some initial tensions between the Rouse company and its founder's new creation.[21] Gradually, however, the two worked out a division of labour. In its early years, therefore, the EDC came to concentrate its efforts on creating festival marketplaces in several smaller American cities. The first was Waterside in Norfolk, Virginia, the first phase of which opened in 1983.[22] Others followed in succeeding years at Portside in Toledo, Ohio; 6th Street Marketplace in Richmond, Virginia; Water Street in Flint, Michigan and McCamley Place in Battle Creek, Michigan.[23] To say the least, however, none of these examples repeated the runaway successes of Boston or Baltimore. The Norfolk example managed eventually to achieve some sustained prosperity but the others, after initial interest, proved disastrous and were soon closed.

One of the key reasons was that the cities concerned were simply too small to provide sustained business for these ill-fated developments. This limited the local market, particularly since these cities' economies were also very weak. Flint, for example, had been shattered by closures in the automobile industry and had 24 per cent unemployment at the time Water Street opened. In addition, a major lesson of Baltimore was being ignored because all these places lacked the clustering of potent leisure and cultural attractors that might have drawn visitors from farther afield. Here Baltimore had a major advantage because the development of *its* critical mass of cultural urbanism had occurred during a period when direct federal funding for city or state projects or to underwrite private projects had been relatively easy to obtain. Ronald Reagan's presidency from 1980 had seen a sharp curtailment of funding from Washington, placing what proved to be an excessive reliance on private funds. Even some of the Rouse Company's ostensibly less risky projects began to experience difficulties, especially as property boom turned to slump in the later 1980s.

Globalising the Model

One consequence was to push EDC increasingly to seek opportunities in big cities outside the United States, especially where public funds were still available properly to prime the pump for development. One of the first and closest of these involvements came at the Darling Harbour in Sydney. In 1983 Thomas Hayson, an Australian developer who owned some property in the area, had learnt about

---

21 Olsen, pp. 315–320.

22 www.ereserve.com/spec.html (accessed 7.6.05); Olsen, pp. 307–309.

23 Olsen, pp. 327–330; Breen and Rigby, 1994, pp. 149–151.

Figure 15.2   Thomas Hayson (left) shows James Rouse work underway on the Darling Harbour, Sydney

Baltimore initially from a newspaper article[24] followed by a visit to the city where he made contact with Millspaugh and Rouse. Hayson was struck by the remarkable similarities between Darling Harbour and Baltimore. At the time the New South Wales government was actively contemplating a major redevelopment of the redundant docks in time to celebrate the bicentennial of European settlement in Australia in 1988. Private developer interest was initially tepid and Hayson found himself virtually alone in making an early commitment to help implement the plans of the new public authority which was created in 1984 to expedite the project.[25] He brought Millspaugh and subsequently Rouse himself to Sydney, retained EDC on a long term basis and introduced them to the major decision makers.[26]

EDC's chief planner, Mort Hoppenfeld, played a key role and was retained for a time by the new Darling Habour Authority to develop the master plan for the new area. This also helped to ensure that the similarity of settings with Baltimore was also reproduced in the provisions of the plan. Other important hands were also involved but this does not alter the fact that Baltimore was, in a very direct sense, the model. The essential elements were indeed very familiar, with a cluster of key attractions: Aquarium, Convention and Exhibition Centre, Maritime Museum, IMAX cinema

---

24   Kevin Perkins, *Dare to Dream: The Life and Times of a Proud Australian* (Sydney: Golden Wattle, 2001), pp. 192–221.

25   Barry Young, Darling Harbour: A New City Precinct, in G. Peter Webber (ed.), *The Design of Sydney: Three Decades of Change in the City Centre* (Sydney: Law Book Co., 1988), pp. 190–213.

26   Perkins, pp.229–298; Olsen, pp. 333–335.

Figure 15.3 Hayson's completed Harbourside development at Darling
Harbour. It was closely modelled on Rouse's Harborplace in
Baltimore.

(added later), attractive public open space and festival marketplace, Harbourside,
developed by Hayson with an EDC stake.

There were also a few individual touches and the development was executed to
a very high standard, not least in the public spaces. Yet Inner Harbor's main creators
were purveying a tested formula that could rapidly be applied within the tight
deadline of of a 1988 opening date. Despite criticisms, they provided a powerful
external legitimacy for what was, at the time, a controversial project, though one
which has proved a major addition to Sydney's leisure and tourist resources.

EDC was also involved in other projects with Hayson, who actually took a stake
in the Rouse offshoot. They collaborated in advising Hayson at Manly Quay, also on
Sydney Harbour.[27] EDC also became involved in Merlin, a company which Hayson
established with UK investors, and undertook work on projects in three British
cities, Birmingham, Glasgow and Manchester.[28] Only the last of these proposed
developments, of the Great Northern Warehouse area, actually proceeded[29]. The
others fell victim to the property recession in the 1990s. However the links with
Rouse and Baltimore can still be glimpsed in the others, especially so in the canalside
development at Brindley Place in Birmingham, even though it was completed by
other hands. This is the most completely Baltimore-like clustering of attractions of
any British city.[30] Another later, more direct, UK connection was forged in Belfast

---

27   www.ereserve.com/ent.html (accessed 07.06.05).
28   Perkins, pp. 299–306.
29   http://www.manchesterupdate.org.uk/article45.htm (accessed 22.06.05)
30   http://www.brindleyplace.com/ (accessed 22.06.06).

Figure 15.4 Port Vell, Barcelona. Rouse's Enterprise Development Corporation played a key role in formulating the development concept.

from 1991 when EDC became part of a consortium for a mixed development with several Baltimore elements on a prominent waterfront site at Lanyon Place.

Though the English-speaking world was the most obvious territory within which Rouse's brand of the Baltimore model could spread, it was certainly not linguistically corralled. Another early and significant international involvement of the EDC was advising on the redevelopment of Port Vell in Barcelona.[31] In the mid-1980s, it was retained to prepare a master plan and strategy for this area, involving the creation of a Baltimore-style 'fun city' on the Moll d'Espanya and a Cruise Ship terminal. The development was completed in 1992 by the port authority, with familiar Baltimore elements such as an aquarium, IMAX cinema, maritime museum and festival-style shopping mall with food court and night clubs.

The dynamics of the connections with Baltimore cannot be established as precisely as they can for Sydney. Yet we can note that Barcelona's mayor during this period, Pasqual Maragall, architect of the city's recent rise, was actually in the Maryland city, at the Johns Hopkins University, just when 'Baltimore-mania' was moving to its height in the early 1980s.[32] Indeed, it was actually from there he was called back to his native city to participate in its newly democratised politics. Despite the growing

31 Han Meyer, *City and Port: Transformation of Port Cities London, Barcelona, New York, Rotterdam* (Utrecht: International Books, 1999), pp. 158–9; www.ereserve.com/spec. html (accessed 7.6.05).

32 Donald McNeill, *Urban Change and the European Left : Tales from the New Barcelona* (London: Routledge, 1999), p. 87.

misgivings of his friend at Johns Hopkins, David Harvey, it would be surprising if he had not brought back a sense of what Baltimore and Rouse had to offer[33].

Another important location on the 'Rouse trail' was Osaka, Japan's second city[34]. In 1988 the city and port authority together formed a development corporation to redevelop the old port. EDC supplied comprehensive development advice and assistance, representing an unprecedented level of co-operation at that time between an American development company and a Japanese public authority. The result was the Tempozan Harbor Village which opened in 1991 offering the familiar mix of Aquarium, IMAX cinema, art museum, festival marketplace and other attractions.

EDC also undertook many other international commissions that showed some degree of continuities with Baltimore Typically (though not invariably) these were waterfront mixed-use and predominantly leisure/retailing developments, or contained some of these elements. Apart from places already mentioned, the EDC website records its work in Melbourne, Izmir, Cairo, Shanghai, Freeport and Port Columbus (Bahamas), Kobe, Sakai (Japan), Rotterdam, Malaga, Rio di Janeiro, Warsaw, Puerto La Cruz and Porlamar (Venezuela), Zagreb, Kuwait City, Cancun (Mexico), Guanica (Puerto Rico) and Bridgetown (Barbados). Nor does this list appear to be comprehensive.[35] For example, there is also evidence of other significant involvements, notably in Singapore and Cape Town.[36] There were also many American commissions and in several cases a financial stake in developments. These tend to show showing increasing diversification from the 'pure' Baltimore model. However an unusually late festival marketplace was at Aloha Tower in Honolulu (opened 1994), one of the few American examples to be built after the failures of the late 1980s.[37] It formed part of a much larger mixed-use waterfront development.

Not all of these commissions, especially the international ones, directly resulted in developments. One reason was that EDC was often giving advice about evolving broad development concepts rather than undertaking a detailed design and management exercise. This was the case, for example, for EDC's advice about the Kop van Zuid in Rotterdam. In other cases, developments fell victim to the uncertainties or fluctuations of the property market (such as the proposed Mall of Egypt in Cairo). But the important point is that everywhere EDC was playing an important role in spreading a common discourse about the future of cities. In effect, it was tilting the global planning and development agenda towards the particular brand of post-industrial urbanism that was associated with Baltimore.

---

33   David Harvey, *The Condition of Post-Modernity* (Cambridge MA: Basil Blackwell, 1989), pp. 66-98; Stephen V. Ward, *Planning the Twentieth-Century City: The advanced capitalist world* (Chichester: Wiley, 2002), pp. 371–372.

34   Olsen, p. 335; www.ereserve.com/spec.html (accessed 07.06.05).

35   www.ereserve.com (accessed 07.06.05).

36   Olsen, p. 335: T.C. Chang, Shirlena Huang and Victor Savage, On the Waterfront: Globalization and Urbanization in Singapore, *Urban Geography*, 25/5 (2004): 413–436; ex info. Peter Wilkinson.

37   www.ereserve.com/spec.html (accessed 07.06.05).

Other Agents of Diffusion

It is important also to stress that the diffusion of the model did not rely entirely on EDC. Some of the professionals who had been associated with Rouse and Millspaugh played an independent part in spreading elements of the model. Thus Benjamin Thompson Associates, architects of Rouse's most successful marketplaces, designed variants of the concept in Guatamala City and Buenos Aires for different clients during the 1990s.[38] They also prepared the master plan for the Custom House Dock in Dublin. There was also a good deal of straightforward borrowing, variously adaptive or uncritical, of ideas from Baltimore by other cities and their planners and developers.[39] Thus knowledge of the city (and also Boston) was particularly influential on activities of many of the British urban development corporations during the 1980s and 1990s. This was particularly so for London, Merseyside, Cardiff and other cities which were undertaking dock regeneration schemes.[40] Collaborative research projects between Universities and other links increased UK awareness of key American developments and helped to reinforce many similarities in urban policy between the two countries during the Reagan-Thatcher era.[41]

In turn cities which had been influenced by Baltimore contributed to spreading the model, or at least their own variants of it, further afield. For example, the city of Barcelona devoted much effort to promoting internationally itself as an urban planning and design model.[42] There was, of course, much in its planning that went beyond Baltimore, especially the new beach waterfronts created during the 1990s. Yet the most Baltimore-like part of Barcelona, Port Vell, became a partial model for the Puerto Madero in Buenos Aires[43]. Under a formal co-operation agreement between the two cities, architects from the Catalonian capital undertook much of the early planning of this area during the late 1980s. Another example of this tendency has been the international consultancy activity of the eponymous development company for the Victoria and Alfred Waterfront in Cape Town.[44] Set up in 1988, its own planning and development was much influenced by the Baltimore experience. During the 1990s it began to undertake commissions in a variety of places including

---

38 www.bta-architects.com/c/portfolio_.html (accessed 17.06.05).

39 Ann Breen and Dick Rigby, *The New Waterfront: A Worldwide Urban Success Story,* (London: Thames and Hudson, 1996).

40 Robert Imrie and Huw Thomas (eds), *British Urban Policy: An evaluation of the urban development corporations,* 2nd edn, (London: Sage, 1999); Duncan Syme, Born Again: The Resurrection of Cardiff Bay, *Locum Destination Review,* 2 (2000): reproduced at http://www.locum-destination.com/pdf/LDR2Born_again.pdf (accessed 23.06.05).

41 For example, Barnekov, Boyle and Rich.

42 Francisco-Javier Monclús, The Barcelona model: an original formula? From 'reconstruction' to strategic urban projects (1979–2004), *Planning Perspectives,* 18/4 (2003): pp. 399–421.

43 www.puertomadero.com/i_planes11.cfm (accessed 17.06.05).

44   www.waterfront.co.za/profile/external_projects/intexpertise.php   (accessed 17.06.05).

Libreville (Gabon), St Louis (Mauritius); Lagos, the Black Sea resort of Gelendzhik, (Russia) as well as Portsmouth (UK) and, coincidentally, about aspects of the development of Port Vell in Barcelona.

Reflections on the Baltimore Model

This last example shows how increasingly interwoven the various strands of influence that originate in Baltimore have now become. The model, in other words has become increasingly globalised, spread by many hands to all parts of the world. Though such places are located within specific cities with a distinctiveness that comes from their physical setting and the character of their buildings (especially so if they predate the renewal process), their essential character has become less and less local.[45] Their success tends to depend more on tourists than on local populations. The outlets the festival marketplaces once provided for local traders have often now become more standardised and predictable (though this depends on rental levels and management policies). And the wider economic impacts of the 'cities are fun' model have been disappointing.

No city exemplifies this more than Baltimore itself, where the workforce of the Inner Harbor and Charles Center is a microcosm of the dual labour market of the wider metropolitan area. On the one hand it provides high paid employment for a largely suburban-dwelling middle class population. On the other the much lower paid jobs of cleaning, security and retailing tend to be taken more by city dwellers. And, as we have seen, a wide section of Baltimore's population is unable to take even these. The acuteness of the city's current problems, in many respects the worst of any American city, raises the question as to whether the Baltimore model may even have been a contributory element. There are certainly other powerful factors, notably a fiscal base that is unusually narrow for an American city, exacerbating the effects of city-suburb polarisation.[46] Nor can drug abuse or the particular problems of gun crime, especially amongst young black males, be attributed solely to economic causes. But there is perhaps a sense in which the perceived early successes of the Inner Harbor have allowed wider opinion to see it as a city 'on the mend'.

The wider truth is also that, over two decades on, the model of cultural urbanism and leisure consumption which the Maryland city once represented has moved on, detached from the place itself. Many other cities, including those discussed in this book, have now mined the seam of urban opportunity that Baltimore discovered almost by accident. Inevitably most of them have tried to improve on the original and have added their own refinements. The newer attractions have become more

---

45   For example, Maarten A. Hajer, 'Rotterdam: Re-designing the Public Domain', in Franco Bianchini and Michael Parkinson (eds), *Cultural Policy and Urban Regeneration: The West European Experience* (Manchester: Manchester University Press, 1993), pp. 48–72, especially pp. 62–64.

46   David Rusk, *Baltimore Unbound: A Strategy for Regional Renewal* (Baltimore: Abel Foundation, 1996).

spectacular than their predecessors with architecture ever more iconic. Inevitably, these newer attempts appear to offer more to would be emulators. And, perhaps more pertinently, many of them have been backed by greater public spending and more helpful metropolitan policies than has benighted Baltimore. Yet the city's long term impact cannot be gainsaid. This, the first realisation of James Rouse's dictum that 'cities are fun' in a troubled city of the industrial era, has now become absorbed into the global repertoire of urban policies. But making sure that cities really are fun for all their inhabitants remains a distant goal.

Index

Publications are indicated by italic type. Tables and figures are indicated by **bold** page numbers.